161 031123 5

C000112828

wo week

n

Please return on or before the last
date stamped below.
Charges are made for late return.

COMPLEXITY
IN BIOLOGICAL
INFORMATION
PROCESSING

Novartis Foundation Symposium 239

COMPLEXITY IN BIOLOGICAL INFORMATION PROCESSING

2001

JOHN WILEY & SONS, LTD

Chichester · New York · Weinheim · Brisbane · Singapore · Toronto

Other Wiley Editorial Offices

John Wiley & Sons, Inc., 605 Third Avenue,
New York, NY 10158-0012, USA

WILEY-VCH Verlag GmbH, Pappelallee 3,
D-69469 Weinheim, Germany

John Wiley & Sons Australia, Ltd, 33 Park Road,
Milton, Queensland 4064, Australia

John Wiley & Sons (Asia) Pte Ltd, 2 Clementi Loop #02-01,
Jin Xing Distripark, Singapore 129809

John Wiley & Sons (Canada) Ltd, 22 Worcester Road,
Rexdale, Ontario M9W 1L1, Canada

Novartis Foundation Symposium 239
viii+240 pages, 45 figures, 0 tables

Library of Congress Cataloging-in-Publication Data

Complexity in biological information processing / [editors, Gregory Bock and Jamie Goode].
p. cm. – (Novartis Foundation symposium ; 239)
Includes bibliographical references.
ISBN 0-471-49832-7 (alk. paper)
1. Biological control systems. 2. Bioinformatics. 3. Cellular signal transduction. 4.
Genetic regulation. I. Bock, Gregory.   II. Goode, Jamie.   III. Symposium on Complexity
in Biological Information (2000 : Berlin, Germany)   IV. Series.
QH508.C66 2001
571.7–dc21   2001033421

British Library Cataloguing in Publication Data

A catalogue record for this book is available from the British Library

ISBN 0 471 49832 7

Typeset in $10\frac{1}{2}$ on $12\frac{1}{2}$ pt Garamond by Dobbie Typesetting Limited, Tavistock, Devon.
Printed and bound in Great Britain by Biddles Ltd, Guildford and King's Lynn.
This book is printed on acid-free paper responsibly manufactured from sustainable forestry,
in which at least two trees are planted for each one used for paper production.

Contents

Participants

Ad Aertsen   Department of Neurobiology and Biophysics, Institute of Biology III, Albert-Ludwigs-University, Schänzlestrasse 1, D-79104 Freiburg, Germany

Michael Berridge   The Babraham Institute, Laboratory of Molecular Signalling, Babraham Hall, Babraham, Cambridge CB2 4AT, UK

Georg Brabant   Computational Endocrinology Group, Department of Clinical Endocrinology, Medical School Hanover, Carl-Neuberg-Str. 1, D-30625 Hanover, Germany

Sydney Brenner   The Molecular Sciences Institute, 2168 Shattuck Avenue, 2nd Floor, Berkeley, CA 94704, USA

Ricardo E. Dolmetsch   Department of Neurobiology and Section of Neuroscience, Harvard Medical School and Children's Hospital, 300 Longwood Avenue, Boston, MA 02115, USA

Gregor Eichele   Max-Planck-Institut für Experimentelle Endokrinologie, Feodor-Lynen-Str. 7, Hanover, D-30625, Germany

R. Douglas Fields   Neurocytology and Physiology Unit, National Institutes of Health, NICHD, Building 49, Room 5A-38, Bethesda, MD 20892, USA

Thomas Gudermann   Department of Pharmacology and Toxicology, Philipps-University Marburg, Karl-von-Frisch-Str. 1, D-35033 Marburg, Germany

Thomas Hofmann (*Novartis Foundation Bursar*)   Freie Universität Berlin, Institut für Pharmakologie, Thielallee 67-73, D-14195 Berlin, Germany

Ravi Iyengar   Department of Pharmacology, Mount Sinai School of Medicine, New York, NY 10029, USA

C. Ronald Kahn   Joslin Diabetes Center, Research Division, Department of Medicine-BWH, Harvard Medical School, Boston, MA 02215, USA

Simon Laughlin   Department of Zoology, University of Cambridge, Downing Street, Cambridge CB2 3EJ, UK

Denis Noble   University Laboratory of Physiology, University of Oxford, Parks Road, Oxford OX1 3PT, UK

Tullio Pozzan   Department of Experimental Biomedical Sciences, University of Padova, Via Trieste 75, 35121 Padova, Italy

Klaus Prank   Research and Development, BIOBASE Biological Databases/ Biologische Datenbanken GmbH, Mascheroder Weg 1b, D-38124 Braunschweig, Germany

Christof Schöfl   Computational Endocrinology Group, Department of Clinical Endocrinology, Medical School Hanover, Carl-Neuberg-Str. 1, D-30625 Hanover, Germany

Günter Schultz   Freie Universität Berlin, Institut für Pharmakologie, Thielallee 69-73, D-14195 Berlin, Germany

Lee Segel   Department of Computer Science and Applied Mathematics, Weizmann Institute of Science, Rehovot 76100, Israel

Terrence Sejnowski (*Chair*)   Computational Neurobiology Laboratory, Salk Institute for Biological Studies, 10010 North Torrey Pines Road, La Jolla, CA 92037-1099, USA

Matthias von Herrath   Departments of Neuropharmacology and Immunology, Division of Virology, The Scripps Research Institute, 10550 North Torrey Pines Road, IMM-6, La Jolla, CA 92037, USA

Introduction

Terrence Sejnowski

Computational Neurobiology Laboratory, Salk Institute for Biological Sciences, 10010 North Torrey Pines Road, La Jolla, CA 92037-1099, USA

I am looking forward, over the next two days, to exploring in depth this exciting and emerging area of biological complexity. It was Dobzhansky who once said that nothing in biology makes sense except in the light of evolution, and this is certainly true of biological complexity. In some ways, complexity is something that many biologists try to avoid. After all, it is a lot easier to study a simple subject than a complex one. But by being good reductionists — taking apart complex creatures and mechanisms into their component parts — we are left at the end with the problem of putting them back together. This is something I learned as a child when I took apart my alarm clock, discovering it didn't go back together nearly as easily as it came apart. What is emerging, and what has given us the opportunity for this meeting, is the fact that over the last few years there has been a confluence of advances in many different areas of biology and computer science which make this a unique moment in history. It is the first time that we have had the tools to actually put back together the many pieces that we have very laboriously and expensively discovered. In a sense, we are at the very beginning of this process of integrating knowledge that is spread out over many different fields. And each participant here is a carefully selected representative of a particular sub-area of biology.

In real estate there is a well known saying that there are three important criteria in valuing a property: location, location and location. In attempting to identify a theme to integrate the different papers we will be hearing in this symposium, it occurred to me that, likewise, there are three important threads: networks, networks and networks. We will be hearing about gene networks, cell signalling networks and neural networks. In each of these cases there is a dynamical system with many interacting parts and many different timescales. The problem is coming to grips with the complexity that emerges from those dynamics. These are not separate networks: I don't want to give the impression that we are dealing with compartmentalized systems, because all these networks ultimately are going to be integrated together.

One other constraint we must keep in mind is that ultimately it is behaviour that is being selected for by evolution. Although we are going to be focusing on these

1

details and mechanisms, we hope to gain an understanding of the behaviour of whole organisms. How is it, for example, that the fly is able to survive autonomously in an uncertain world, where the conditions under which food can be found or under which mating can take place are highly variable? And how has the fly done so well at this with such a modest set of around 100 000 neurons in the fly brain? We will hear from Simon Laughlin that one of the important constraints is energetics.

I have a list of questions that can serve as themes for our discussion. I want to keep these in the background and perhaps return to them at the end in our final discussion session. First, are there any general principles that will cut across all the different areas we are addressing? These principles might be conceptual, mathematical or evolutionary. Second, what constraints are there? Evolution occurred for many of these creatures under conditions that we do not fully understand. We don't know what prebiotic conditions were like on the surface of the earth, and this is partly why this is such a difficult subject to study experimentally. The only fossil traces of the early creatures are a few forms preserved in rock. What we would really like to know is the history, and there is apparently an opportunity in studying the DNA of many creatures to look at the past in terms of the historical record that has been left behind, preserved in stretched of DNA. But the real question in my mind concerns the constraints that are imposed on any living entity by energy consumption, information processing and speed of processing. In each of our areas, if we come up with a list of the constraints that we know are important, we may find some commonality. The third question is, how do we make progress from here? In particular, what new techniques do we need in order to get the information necessary for progress? I am a firm believer in the idea that major progress in biology requires the development of new techniques and also the speeding-up of existing techniques. This is true in all areas of science, but is especially relevant in biology, where the impact of techniques for sequencing DNA, for example, has been immense. It was recently announced that the sequence of the human genome is now virtually complete. This will be an amazingly powerful tool that we will have over the next 10 years. As we ask a particular question we will be able to go to a database and come up with answers based on homology and similarities across species. Who would have guessed even 10 years ago that all of the segmented creatures and vertebrates have a common body plan based around the *Hox* family of genes? This is something that most of the developmental biologists missed. They didn't appreciate how similar these mechanisms were in different organisms, until it was made obvious by genetic techniques. Another technique that will provide us with the ability both to do experiments and collect massive amounts of data is the use of gene microarrays. It is now possible to test for tens of thousands of genes in parallel. We can take advantage of the fact that over the last 50 years, the performance of computers, both in terms of memory and processing power, has been rising exponentially. In 1950 computers based on vacuum tubes could do about 1000 operations per second; modern parallel supercomputers are capable of around 10^{13} operations per second. This is going to be of enormous help to us, both in terms of keeping track of information and in performing mathematical models. Imaging techniques are also extremely powerful. Using various dyes, it is possible to get a dynamic picture of cell signalling within cells. These are very powerful techniques for understanding the actual signals, where they occur and how fast they occur. Please keep in mind over the next few days that we need new techniques and new ways of probing cells. We need to have new ways of taking advantage of older techniques for manipulating cells and the ability to take into account the complexity of all the interactions within the cell, to develop a language for understanding the significance of all these interactions.

I very much look forward to the papers and discussions that are to follow. Although it will be a real challenge for us to understand each other, each of us coming from our own particular field, it will be well worth the effort.

Functional modules in biological signalling networks

Upinder S. Bhalla and *Ravi Iyengar[1]

*National Centre for Biological Sciences, Bangalore, India and *Department of Pharmacology, Box 1215, Mount Sinai School of Medicine, One Gustave Levy Place, New York, NY 10029, USA*

Abstract. Signalling pathways carry information from the outside of the cell to cellular machinery capable of producing biochemical or physiological responses. Although linear signalling plays an important role in biological regulation, signalling pathways are often interconnected to form networks. We have used computational analysis to study emergent properties of simple networks that consist of up to four pathways, We find that when one pathway gates signal flow through other pathways which produce physiological responses, gating results in signal prolongation such that the signal may be consolidated into a physiological response. When two pathways combine to form a feedback loop such feedback loops can exhibit bistability. Negative regulators of the loop can serve as the locus for flexibility whereby the system has the capability of switching states or functioning as a proportional read-out system. Networks where bistable feedback loops are connected to gates can lead to persistent signal activation at distal locations. These emergent properties indicate system analysis of signalling networks may be useful in understanding higher-order biological functions.

2001 Complexity in biological information processing. Wiley, Chichester (Novartis Foundation Symposium 239) p 4–15

Complexity is a defining feature of signal flow through biochemical signalling networks (Weng et al 1999). This complexity arises from a multiplicity of signalling molecules, isoforms, interactions and compartmentalization. This leads to significant practical problems in understanding signalling networks. On the one hand, the common 'block diagram' description of signalling pathways is lacking in quantitative detail. On the other, a listing of all the rate constants in a pathway (assuming they are available) also does not convey much understanding. Depending on the signalling context, it is very likely that details such as the fine balance between rates of action of competing pathways, or the timing of series of

---

[1]The chapter was presented at the symposium by Ravi Iyengar to whom correspondence should be addressed.

4

reactions are critical determinants of the outcome of signal inputs. Computational analysis is poised to bridge this divide between crude abstractions and raw data. In this paper we will discuss the emergence of more useful functional concepts from the molecular building blocks, and consider how these might behave in combination.

We have developed experimentally constrained models of individual enzyme regulation and signalling pathways described in terms of molecular interactions and enzymatic reactions. Biochemical data from the literature were used to work out mechanisms and specify rate constants and concentrations. These values were entered and managed using the Kinetikit interface for modelling signalling pathways within the GENESIS simulator. Simulations have been carried out on PCs running Linux. Modelling and parameterization methods have been previously described (Bhalla 1998, 2000)

We have examined four major protein kinase pathways and their regulators: protein kinase C (PKC); the mitogen-activated protein kinase (MAPK); protein kinase A (PKA); and the Ca^{2+}–calmodulin-activated protein kinase type II (CaMKII). Reaction details have been previously reported (Bhalla & Iyengar 1999). Figure 1 describes in block diagram form the molecular interactions and inputs into a network containing these four protein kinases. The block diagram in Figure 1 is clearly complex, and the underlying reaction details are even more so. How then can the functioning of such a system be understood? Our computational analyses suggest a set of functional modules which capture the essential behaviour of the system and also facilitate prediction of responses. The behaviour of each module is strongly dependent on the details of the reaction kinetics and mechanisms, and is often context-dependent. These details are readily examined through simulations but tend to become obscured by block-diagram representations. One of the goals of describing the system in terms of functional modules is to provide a conceptual tool for examining signal flow, which is nevertheless based on the molecular details. Some of the key signalling functions we observe are gating, bistable feedback loops, coincidence detectors, and regulatory inputs.

Gating

Gating occurs when one signalling pathway enables or disables signal flow along another. In the present system, this is illustrated by the action of PKA on CaMKII responses to Ca^{2+} influx. Biochemical experiments show that this regulation occurs when PKA phosphorylates the inhibitory domain of the protein phosphatase, PP1. This phosphorylation turns off PP1 by activating the inhibitor. This interaction plays a gating role because PP1 rapidly reverses the autophosphorylation of CaMKII and hence prevents long-term activation of CaMKII and consequently

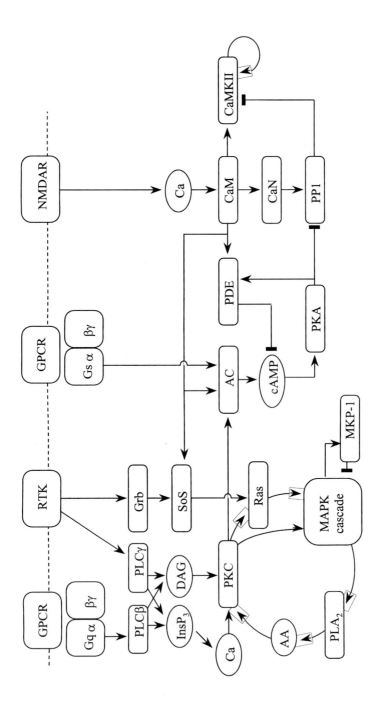

FIG. 1. A signalling network that contains the four major protein kinases (PKA, PKC, MAPK 1, 2 and CaMKII). The inputs to the kinases and the interconnections are shown. The outputs for the kinases are not shown. Each protein kinase can phosphorylate and regulate multiple targets. Detailed analysis of this network within the context of the post-synaptic region of the CA1 neuron is described in Bhalla & Iyengar (1999).

long-term potentiation. Supporting evidence for this interaction comes from experiments on long-term potentiation where activation of the cAMP pathway was a prerequisite for synaptic change (Blitzer et al 1995, 1998). As described in these papers, PKA activation gates CaMKII signalling by regulating the inhibitory process that deactivates the persistently activated CaMKII.

Coincidence detectors

There is some overlap between the concept of gating and that of coincidence detection. The former implies that one pathway enables or disables another. The latter suggests that two distinct signal inputs must arrive simultaneously for full activation. The requirement of timing is a distinguishing feature between the two. Coincidence detectors typically involve situations in which both inputs are transient, whereas gating processes are usually prolonged. At least two coincidence detectors are active in the case of the pathways considered in Figure 1. First, PKC is activated to some extent by Ca^{2+} and diacylglycerol (DAG) individually, but there is a strong synergistic interaction such that simultaneous arrival of both signals produces a response that is much greater than the additive response (Nishizuka 1992). Ca^{2+} signals arrive in various ways, notably through ion channels and by release from intracellular stores. DAG is produced by the action of phospholipase C (PLC) β and γ, which also mediate Ca^{2+} release from intracellular stores via inositol-1,4,5-trisphosphate (InsP$_3$). At synapses the coincident activation of these two pathways occurs through strong stimulation resulting in glutamate release. As described elsewhere, an important step in synaptic change occurs when the NMDA receptor opens on an already depolarized synapse, leading to Ca^{2+} influx (Bliss & Collingridge 1993). Simultaneously, the metabotropic glutamate receptor (mGluR) is also activated, turning on PLC. The PLC cleaves phosphatidylinositol-4,5-bisphosphate (PIP$_2$) into InsP$_3$ and DAG. The coincident arrival of DAG and Ca^{2+} strongly activates PKC. A second important coincidence detection system is the Ras pathway, acting through the MAPK cascade in this model (Fig. 1). Ras is activated by several inputs, but for our purposes it is interesting to note that simultaneous receptor tyrosine kinase (RTK) as well as G protein-coupled receptor input can act synergistically to turn on Ras. Due to the strongly non-linear nature of MAPK responses, coincident activation produces responses that are much greater than either individual pathway.

Bistable feedback loops

Bistable feedback loops are among the most interesting functional modules in signalling. In this system, such a loop is formed by the successive activation of

MAPK by PKC, of PLA2 by MAPK, and the formation of arachadonic acid (AA) by PLA2 and the activation of PKC by AA (Fig. 1). Bistable systems can store information. This occurs because brief input signals can 'set' the feedback loop into a state of high activity, which will persist even after the input has been withdrawn. Thus the information of the previous occurrence of a stimulus is stored in the feedback loop. We have previously shown that transient synaptic input can lead to prolonged activation of this biochemical bistable loop (Bhalla & Iyengar 1999). The system also exhibits sharp thresholds for stimuli. Feedback loops have the potential to act as biochemical 'engines' driving several emergent signalling phenomena.

Regulation of feedback

The range of operation of this feedback circuit is still further extended by regulatory inputs. These are worth considering as distinct functional modules because of the additional functions they confer upon the basic feedback loop. In our system, one such regulatory signal is provided by MAPK phosphatase 1 (MKP-1). MKP-1 itself is synthesised in response to MAPK activation. MKP-1 and another inhibitory regulator of the MAPK cascade, PP2A, can each regulate the mode of action of the feedback system. These modes include linear responses with variable gain; 'timer switching' which turns on in response to brief stimuli but turns off after delays ranging from tens of minutes to over an hour; or as constitutively 'on' or 'off' systems. Furthermore, slow changes in regulator levels can elicit sharp irreversible responses from the feedback circuit in a manifestation of catastrophic transitions (Bhalla & Iyengar 2001).

Modularity and integrated system properties

There is clearly a rich repertoire of functional behaviour displayed by a signalling network. The specific responses in a given biological context are governed by the details of the signalling kinetics and interactions, and are not readily deduced simply from the pathway block diagram. Once one has identified the likely functional modules, it is possible to examine the integrated behaviour of the system from a different viewpoint. The reclassification of the same network in terms of functional modules rather than chemical blocks is shown in Fig. 2. Using such modularity as the basis for analysis, we can begin to understand many of the aspects of system behaviour that tend to defy intuition based on molecular block diagrams. These include:

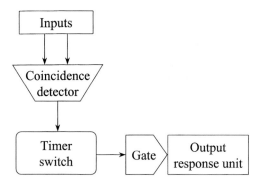

FIG. 2.   The functional modules that comprise the signalling network shown in Fig. 1. In this context the four protein kinases are parts of different modules including the timer switch, the gate and the response unit.

(1)   The feedback loop as a key determinant of overall system responses. In this context the feedback loop acts as a timer switch sensitive to very brief inputs and is capable of maintaining an output for around an hour.

(2)   The presence of a coincidence detector in the inputs to the timer switch. This configuration suggests that simultaneous activation of multiple pathways to activate PKC may be more effective in turning on the switch than individual inputs.

(3)   The output of the timer switch as a feed to a gating module that affects CaMKII function. Weak stimuli will activate CaMKII in a transient manner, since the gate will rapidly shut down its activity. Stronger stimuli open the gate by activating the feedback loop. This provides a mechanism for selective prolongation of CaMKII activity.

The modular organization of the signalling network in Fig. 1 as described above is shown in Fig. 3.

With such a functional outline of the signalling network, one can now return to the biological context to assess the likely implications. In this network, for example, there is a clear suggestion that the termination process for the switch (induction of MKP-1 synthesis by MAPK) may in parallel induce other synaptic proteins. These proteins could therefore integrate into the synapse to 'take over' from the switch at precisely the same time as the switch itself is turned off by MKP-1. The cytoskeletal roles of CaMKII and MAPK suggest further specific possibilities for how these changes might occur in a spatially restricted manner. Experimental reports also support this notion of synaptic 'tagging', in which strong stimuli induce activity in specific synapses and lead to synthesis of new

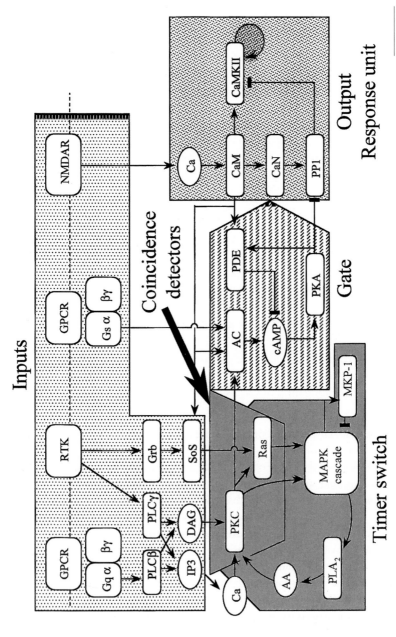

FIG. 3. Abstraction of the signalling network into functional units described in Fig. 2. In this context the four protein kinases are parts of different functional modules. However in reality all of the four protein kinases retain the ability to function as output response units, and may do so for different cellular functions.

proteins, which are selectively taken up at the 'tagged' synapses (Frey & Morris 1997).

Understanding complexity

A key question in performing detailed computational analyses is: does exhaustive detail really lead to a better understanding of the system? It is often felt that detailed models appear to simply map one complex system (interacting molecules) onto an equally complex one (a computer model) without highlighting the underlying principles that define the system. The process of modelling does not support this pessimistic view. Modelling gives one the tools to identify simple conceptual and functional modules from amongst the mass of molecular interactions. This is not merely a matter of grouping a set of molecules and interactions into a new module according to some fixed classification. The configuration as well as the operation of these modules is highly dependent on the specific details of the system, so one cannot simply replace a signalling block diagram with a functional one. For example, the experimental parameters placed our positive feedback loop in a regime where it is most likely to act as a timer switch. Other parameters could readily have made it into a linear responsive element, or even an oscillator (Bhalla & Iyengar 2001). Other feedback loops, comprising of completely different molecules, would exhibit a similar repertoire of properties, with the similar dependence on the exact signalling context. This includes the most intuitively obvious function of a positive feedback loop, signal amplification. The functional description is therefore useful as a level of understanding, and not merely a classification device.

Analysis

Once the system identification has been performed, it is much easier to analyse signal flow in the network in terms of functional entities rather than simply molecular ones. The network we use as an example was reduced to three or four functional elements, whose interactions were rather simple. One could build on this approach by considering a greater number of pathways as well as by acknowledging the presence of additional interactions among the existing ones. For instance, PKA is known to negatively gate the Ras pathway in some biological systems, depending on the isoform of Raf that is present. Our functional network would suggest that this should rapidly turn off the feedback system, perhaps even before it could reach full activity. This would depend on the relative ratios of the isoforms of Raf differentially regulated by PKA. Thus we can define functions of the modules and their interactions in terms of the identity and concentrations of the molecular components within the modules. It is also much easier now to consider the operation of the same functional units in a different context, for example in triggering proliferative responses. Although the inputs and many of the intermediate players are now different, one can experimentally demonstrate responses that are consistent with the presence of a bistable feedback loop in growth-factor stimulated cells (Gibbs et al 1990). The properties of the feedback loop provide a clear basis for thinking about how thresholds are set and sustained responses obtained for this different physiological function.

Biological context

The process of analysing signalling is brought full circle by placing the functional modules back into the biological context to ask what the response might mean for the cell. At this point we would have an opportunity to describe and evaluate events which may have been obscured by the abstraction. In the synaptic context we have numerous potential interactions, not only at the putative signalling end-points in this model (the four kinases), but also at the level of intermediate regulators such as the phospholipases. The essential purpose of the whole exercise, of course, is to advance the state of understanding of the system as a whole with the simultaneous knowledge of the role each individual component and reaction plays in this systems property. The abstract functional description, the detailed simulations, and the experimental data are meant to feed into each other to predict system behaviour in terms of molecular components and interactions and suggest fruitful lines of further investigation.

Acknowledgements

This research in the Iyengar laboratory is supported by NIH grant GM-54508 and grants from NCBS to the Bhalla laboratory.

References

Bhalla US 1998 The network within: signalling pathways. In: Bower JM, Beeman D (eds) The book of GENESIS: exploring realistic neural models with the general neural simulation system. Springer-Verlag, New York, p 169–190
Bhalla US 2000 Simulations of biochemical signalling. In: De Schutter E (ed) Computational neuroscience: realistic modelling for experimentalists. CRC Press, Boca Raton, FL, p 25–48
Bhalla US, Iyengar R 1999 Emergent properties of networks of biological signaling pathways. Science 283:381–387
Bhalla US, Iyengar R 2001 Robustness of a biological feedback loop. Chaos 11:221–226
Bliss TV, Collingridge GL 1993 A synaptic model of memory: long-term potentiation in the hippocampus. Nature 361:31–39
Blitzer RD, Wong T, Nouranifar R, Iyengar R, Landau EM 1995 Postsynaptic cAMP pathway gates early LTP in hippocampal CA1 region. Neuron 15:1403–1414

Blitzer RD, Connor JH, Brown GP et al 1998 Gating of CaMKII by cAMP-regulated protein phosphatase activity during LTP. Science 280:1940–1942

Frey U, Morris RG 1997 Synaptic tagging and long-term potentiation. Nature 385:533–536

Gibbs JB, Marshall MS, Skolnick EM, Dixon RA, Vogel US 1990 Modulation of guanine nucleotides bound to ras in NIH3T3 cells by oncogenes, growth factors, and the GTPase activating protein (GAP). J Biol Chem 265:20437–20442

Nishizuka Y 1992 Intracellular signalling by hydrolysis of phospholipids and activation of protein kinase C. Science 258:607–614

Weng G, Bhalla US, Iyengar R 1999 Complexity in biological signaling systems. Science 284: 92–96

DISCUSSION

Sejnowski: You mentioned long-term potentiation (LTP), which is one of the most controversial issues in neurobiology. Chuck Stevens has evidence for changes occurring in presynaptic terminals, whereas Roger Nicoll sees changes in the postsynaptic side. The biochemical basis of LTP is even more complicated. Mary Kennedy has addressed this issue: why is it that there is so much controversy over LTP (Kennedy 1999)? Are physiologists not doing the experiments properly, or could they be using the wrong model? Physiologists look at signalling in terms of a linear sequence of events: the voltage gates the channel, the channel opens, current flows and this causes an action potential. In other words, there is a nice progression involving a sequence of events that can be followed all the way through to behaviour of the axon, as Hodgkin and Huxley first showed. But could it be that LTP is not like that? Perhaps LTP is much closer to a system such as the Krebs cycle. The diagrams you showed looked more like metabolism to me than an action potential. If this is true, perhaps we are thinking about things in the wrong way.

Iyengar: My collaborator, Manny Landau, was a collaborator with Chuck Stevens back when Chuck was at Yale. In theory, we belong to the presynaptic camp, except that most of our experiments seem to work postsynaptically. We don't want to upset Chuck, but we don't as yet have any data that indicate a presynaptic locus for the functions we study. One of the reasons we conceived the large-scale connections map I described is that many of the same pathways that work postsynaptically also function presynaptically. We are limited by the tools we have. We can easily get things into the postsynaptic neuron, but there is currently no real way of getting stuff into the CA3 neuron and working out the presynaptic signalling network.

Sejnowski: Suppose that we have a system with a whole set of feedback pathways that involves not just the postsynaptic element, but also the presynaptic and even the glial cells. There is a lot of evidence for interactions between all these elements. Also, time scales are important. There is short-term, intermediate and long-term potentiation. Associated with each of these timescales will be a separate biochemistry and set of issues. For example, Eric Kandel and others have shown that for the very longest forms of synaptic plasticity, protein synthesis and gene regulation are necessary. This takes hours.

Iyengar: Indeed. In our large-scale connections map, we have translation coupled here, when in reality in the LTP model translation is after the movement machinery. In the most recent papers, the translation that goes on in LTP seems to be at the dendrites. There is some mechanism that allows this RNA to come and move out to the dendrites, and this is where the real biochemistry happens. One of the focuses that people have is on the Rho–integrin signalling pathway, because this can send signals through MAPK to the nucleus, and at the same time mark the dendrites.

Eichele: What are the contributions of positive and negative feedback loops at the cellular level? In developmental biology feedback regulation is important and can be positive or negative.

Iyengar: It appears that signal consolidation is always required at the cellular level. It could almost be a shifting scale as well. Some key enzyme, in most cases a protein kinase, needs to be activated at a certain level for a certain length of time. These positive feedback loops allow this to happen. In the case of the MAPK pathway I showed, going back and activating PKC allows MAPK to stay active for much longer than the initial EGF signal. In the case of CaMKII, it is the autophosphorylation that allows CaM kinase to stay active for an extended period after the initial Ca^{2+} signal has passed through. Clearly, regulation of the kinase/phosphatase balance is going to be important for signal consolidation. What is not clear in my own mind is whether the timescales over which the signal consolidation occurs are different for different phenomena. My initial guess is that they will be different. The initial MAPK marking in the dendrites, which is a good model for polarity, is going to be very rapid, while the amount of MAPK activation required for gene expression is going to take much longer. This may account for why, if you don't keep it active for long enough, the system depontentiates, but if you go past this 30–40 minute barrier, LTP can be sustained.

Fields: I have some questions relating to the constraints. The general principle of your approach is one based upon kinetic modelling. The assumption is that this problem can be modelled using equilibrium reaction kinetics and constants. To what extent is this valid when the cell is in a dynamic state and the stimulus is dynamic, and how well are the concentrations of the reactants and the kinetic constants known in actual cells? A related question is, given the spatiotemporal constraints, how confident can one be in modelling and knowing that one has set up the right system of reactions when some of these reactions, such as phosphatase feedback loops, may only come into play under certain stimulus conditions?

Iyengar: This is a preliminary model. This is all deterministic, whereas in reality half of life is probably stochastic. We need to include stochastic processes. Many scaffolds and anchors are showing up, and one of their roles is to bring reactants together, anchor them and raise their effective concentrations. The model we have been thinking about most is MAPK. With very low stimulations — single boutons — there is MAPK activated at the dendrites. The model here is that as the MAPK moves up towards the nucleus, it marks the tracks. This is what will give you the 'activated dendrite' that knows that your protein has to come through. This model process is most likely to be stochastic. The problem computationally is not so much dealing with stochastic processes or deterministic processes, but dealing with the boundaries between these processes. Consider that you have 100 molecules of MAPK, and given the temporal aspects of this reaction 40 of them behave stochastically. The question arises as to when these 40 molecules can be integrated back into the deterministic part of the reaction. We don't have real solutions for this issue. Space is another issue we haven't dealt with seriously. With the MAPK model there is one clear compartment between the cytoplasm and the nucleus. MAPK is phosphorylated and goes into the nucleus, but it is clamped there until it is dephosphorylated. If we can map the nuclear phosphatases we can count what is in the nucleus, and see what those rates are.

Brenner: Roughly how many molecules are present?

Iyengar: In the last model I showed you, without taking into account the isoforms, there are about 400 molecules in this connections map.

Brenner: Is this a measured number?

Iyengar: This comes from the actual number of known components. The number of 400 is a gross underestimation, because each of these molecules has at least two or three isoforms present in each neuronal cell. Three would be a reasonable guess.

Brenner: So it is in the order of 10^3 molecules.

Reference

Kennedy MB 1999 On beyond LTP. Long-term Potentiation. Learn Mem 6:417–421

Design of immune-based interventions in autoimmunity and viral infections — the need for predictive models that integrate time, dose and classes of immune responses

Matthias G. von Herrath

Division of Virology, The Scripps Research Institute, 10550 North Torrey Pines Road, IMM-6, La Jolla, CA 92037, USA

Abstract. The outcome of both autoimmune reactions and antiviral responses depends on a complex network of multiple components of the immune system. For example, most immune reactions can be viewed as a balance of aggressive and regulatory processes. Thus, a component of the immune system that has beneficial effects in one situation might have detrimental effects elsewhere: organ-specific immunity and autoimmunity are both governed by this paradigm. Additionally, the precise timing and magnitude of an immune response can frequently be more critical than its composition for determining efficacy as well as damage. These issues make the design of immune-based interventions very difficult, because it is frequently impossible to predict the outcome. For example, certain cytokines can either cure or worsen autoimmune processes depending on their dose and timing in relation to the ongoing disease process. Consequently, there is a strong need for models that can predict the outcome of immune-based interventions taking these considerations into account.

2001 Complexity in biological information processing. Wiley, Chichester (Novartis Foundation Symposium 239) p 16–30

We are unravelling the molecular basis of cellular functions, interactions and effector mechanisms of the immune system at an increasingly rapid pace. The 'mainstream' scientific approach is to isolate single components, characterize them *in vitro* and subsequently probe their *in vivo* function by using genetic knockout or over-expressor animal or cellular models. Although this strategy has significantly furthered our understanding, it has also generated inexplicable situations, for example in that the same molecule might appear to have different functions *in vivo* than it exhibits *in vitro*. The causes of these dilemmas are the

16

redundancy in biological pathways, the issue of compartmentalization and the 'Δt' as well as 'Δc', which is the change in factors or concentrations over time that can frequently be as important as their absolute levels. At this point, there is no clear way to introduce these concepts into our predictive modelling systems for the immune system, and therefore many issues have to be resolved empirically or by trial and error. As a consequence, there are many published observations that appear to be contradictory and cannot be reconciled, which generates confusion rather than understanding. The purpose of this article is to illustrate these considerations with practical examples from our work and that of others in the areas of autoimmunity and viral infections. It will become clear that appropriate models that can describe and predict complex systems would be extremely valuable for bringing immune-interventive therapies closer to the clinic and in increasing our understanding of immunobiology.

Autoimmunity

Regulatory versus aggressive classes of immune responses

Our laboratory is interested in understanding the regulation of autoimmunity. Our previous work, and that of others, has shown that the amount of immuno-pathology or tissue injury is determined not only by the magnitude and precise timing of a localized or systemic immune process, but also to a large extent by the components or the class(es) of responses it encompasses (Homann et al 1999, Itoh et al 1999, Seddon & Mason 1999, von Herrath 1998, von Herrath et al 1995a, 1996). Thus, each immune or autoimmune reaction has aggressive and regulatory components that balance each other out, and these have a strong effect on the duration or magnitude of the response and resulting tissue injury (Liblau et al 1995, Racke et al 1994, Rocken et al 1996, Weiner 1997). In autoimmune diseases, it is possible to take therapeutic advantage of this paradigm and generate autoreactive regulatory cells by targeted immunization with self-antigens. We have shown that such cells can be induced by oral immunization (Homann et al 1999), DNA vaccines (Coon et al 1999) and peripheral immunization. These cells are able to selectively suppress an ongoing autoimmune reaction, because they are preferentially retained in the draining lymph node closest to the target organ where they exert their regulatory function (see Fig. 1). It is clear that certain 'regulatory' cytokines are favourable for autoimmune diabetes in preventing islet destruction, whereas others enhance the pathogenic process. Studies from our lab and others have shown that interleukin 4 has beneficial effects and is required when protecting from autoimmune diabetes by vaccination (Homann et al 1999, King et al 1998). In contrast, induction of interferons generally enhances disease.

Local, 'professional' APC:
Presents self-antigens **A** and
B in inflammatory lesion

Target cell under attack:
Expresses self-antigens **A** and **B**
(β cell, oligodendrocyte)

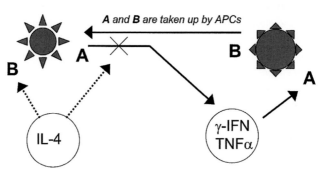

**Autoreactive regulatory Th2
lymphocyte reactive to self-
antigen B:**
Secretes IL-4 (other?) after
recognizing self-antigen **B** on APC,
which prevents APC from activating
Th1 cells reactive to self antigen **A**

**Autoreactive CTL or inflammatory
Th1 lymphocyte reactive to self-
antigen A:**
Attacks some target cells, but
depends on local APCs to present
antigen **A** for expansion/activation

FIG. 1.   Regulation of autoimmunity as a function of auto-aggressive and autoreactive
regulatory responses—the concept of bystander suppression. APC, antigen-presenting cell;
CTL, cytotoxic T lymphocyte; IFN, interferon; IL, interleukin; Th, helper T lymphocyte.

Current studies are dissecting the precise mechanism(s) of action for regulatory,
autoreactive cells (modulation of antigen presenting cells, cytokines/chemokines,
cell contact inhibition) as well as the requirements for their induction (endogenous
autoreactive regulatory T cell repertoire; route and dose of external antigen
administration; expression level and involvement of the endogenous self-antigen
in the autoimmune process). Paradigms developed from these studies will be
useful in suppressing autoimmune diseases very selectively. In general,
autoreactive regulatory lymphocytes are thought to act as 'bystander
suppressors' (Homann et al 1999, Racke et al 1994). This means that an auto-
aggressive process initiated in response to an auto-antigen 'A' can be modulated
by auto-regulatory lymphocytes specific to another auto-antigen 'B' that is also
specific for the targeted organ, and is released and presented to the immune
system by antigen presenting cells after destruction has been initiated by auto-
aggressive cells specific for 'A' (Fig. 1).

Opposing effects of the same cytokine on an ongoing autoimmune process
— levels as well as timing are key issues

Recent findings using cytokine overexpressor or knockout mice have yielded conflicting results for the function of several cytokines in either preventing or enhancing autoimmune diabetes (Cope et al 1997a,b). Key factors influencing the role of a given cytokine in disease are the level, timing in relation to the disease process, and the rate of increase. For example, interferon γ can enhance inflammation and actively participate in β cell destruction (Lee et al 1995, von Herrath & Oldstone 1997), but can also abrogate disease by increasing islet cell regeneration or by augmenting activation-induced cell death (AICD) in autoaggressive lymphocytes (Horwitz et al 1999). Similarly, local expression of interleukin 2 in islets can enhance autoimmunity, but can also abrogate disease by enhancing AICD (von Herrath et al 1995b). Interleukin 10 can have differential effects as well depending on its local 'dose' (Balasa & Sarvetnick 1996, Lee et al 1994, Wogensen et al 1993). Recent observations from our laboratory in a mouse model of virally induced autoimmune diabetes show that production of tumour necrosis factor (TNF)α early or late during the disease process can halt the inflammation leading to diabetes, whereas its expression at the height of islet infiltration enhances incidence and severity of type 1 diabetes (Christen et al 2001). Thus, cytokine levels as well as the time-point of cytokine expression are crucial for defining their function in the disease process and to understanding their role in pathogenesis of autoimmunity.

Since the molecular understanding of immune responses progresses at a very rapid pace, it is frequently impossible to make simple predictions, because the number of molecules and cells involved is too high and their interactive network is too complex. Furthermore, the relative contribution of the different 'players' has to be taken into account as a major factor and this is probably at least one of the reasons why different research teams are frequently reporting seemingly opposing or conflicting results. Such issues might profit from appropriate mathematical or other computer-based modelling systems, which ultimately would allow us to predict more reliably the outcome of interventions for viral infections or autoimmune syndromes. However, before such computer-based models can be developed, we need to have the numerical data to 'feed' into the programs. This has not yet been achieved to a sufficient level. For example, an improved *in vivo* imaging system that permits tracking of specific lymphocytes in animal models and/or humans in a non-invasive way will be instrumental to achieve this.

The goal of immune-based interventions is to preserve stage 2 and prevent its progressing towards the clinical stage 3. Molecules known to be instrumental in this decision are inflammatory and regulatory cytokines, chemokines, adhesion

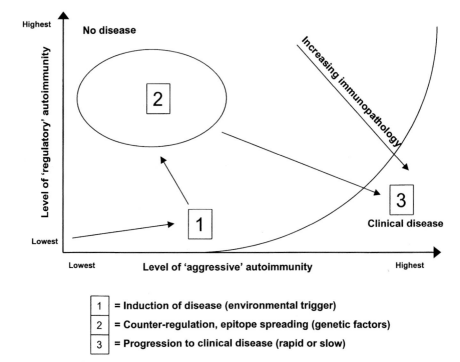

FIG. 2.   Levels of 'aggressive' autoimmunity. The goal of immune-based interventions is to preserve stage 2 and prevent its progressing towards the clinical stage 3. Molecules known to be instrumental in this decision are inflammatory and regulatory cytokines, chemokines, adhesion molecules and the activation profile of autoreactive lymphocytes as well as antigen presenting cells. Many of these molecules can have beneficial or detrimental effects based on the time and level of expression in relation to the ongoing disease process. Due to the complexity of this situation, it has therefore been very difficult to make good predictions about the safety and efficacy of a given approach. Importantly, many of these molecules can be assessed as markers in the peripheral blood and could potentially be used as a basis for a predictive model that would allow rating of the success of immune interventions.

molecules and the activation profile of autoreactive lymphocytes as well as antigen presenting cells. Many of these molecules can have beneficial or detrimental effects based on the time and level of expression in relation to the ongoing disease process. Due to the complexity of this situation, it has therefore been very difficult to make good predictions about the safety and efficacy of a given approach. Importantly, many of these molecules can be assessed as markers in the peripheral blood and could potentially be used as a basis for a predictive model that would allow rating of the success of immune interventions (Fig. 2).

Antiviral immunity

Usually, vaccine-based strategies attempt to enhance immunity to infectious agents (Klavinskis et al 1990). This is generally successful, if the pathogen itself damages host tissues and can be eliminated completely. However, dampening the antiviral response can ameliorate viral immunopathology especially in persistent viral infections but also in some acute situations, where the immune system over-does its 'job' and an intolerable amount of tissue or organ damage is occurring while the infection is cleared (von Herrath et al 1999). This can be achieved with altered or blocking peptides (Bot et al 2000, von Herrath et al 1998) or, more recently, using 'killer' dendritic cells, both of which abrogate antiviral lymphocytes (Matsue et al 1999). Since such immune modulations will curtail the antiviral response, it might be important in some situations that viral replication is suppressed at the same time by using antiviral drug therapy in order to avoid generation of unacceptably high systemic viral titres. Thus, to achieve the desired effect, the immune response has to be suppressed in a very controlled manner and it would be helpful to be able to model/predict the outcome and fine-tune the intervention accordingly.

Similar to the situation in autoimmunity, augmenting or decreasing the antiviral response during an ongoing infection can be either beneficial or detrimental. Many viral infections will not fall neatly into the extreme categories 1–3 indicated in Fig. 3, but instead will be in the 'middle section'. Since the viral load is a function of the efficacy of the immune response and concomitant antiviral drug therapy, the prediction of the outcome of immune dampening or enhancing interventions is complex. It depends on the replication rate of the virus, number of antigen-presenting cells infected, lytic damage of the viral infection to the host cell and the precise kinetics and effector molecules of the antiviral response (cytokines, perforin, FAS, etc.). Many of these factors have been characterized in experimental models and could form the basis for designing appropriate predictive model systems.

Conclusions and future outlook

Immunological processes governing autoimmunity and antiviral responses are too complex to be predicted with methods available to date. Immune modulatory interventions relying on changing the kinetics of an ongoing local or systemic response are currently under evaluation, but would greatly profit from a predictive model that is based on empirical data as well as assessment of the immunological status of a given individual. To obtain such a model, we will need the ability to derive systemic data (non-invasively), for example by determining antigen specific cells as well as antibody levels in the peripheral blood, which has

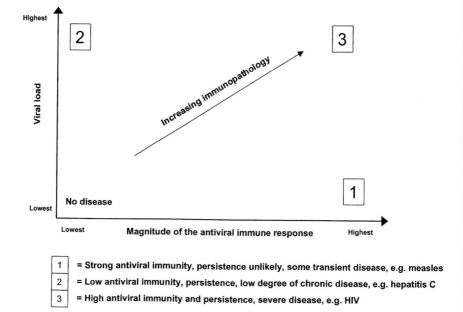

FIG. 3.   Regulation of virally induced immunopathology as a function of viral load and the magnitude of the immune response.

almost become a reality with novel techniques such as MHC tetramers and intracellular fluorescence-activated cell sorting (FACS) analysis. However, the peripheral blood only offers us a tiny and narrow window to the overall immune-status of an individual: dynamic changes and compartmentalization of antigen-specific immune reactions can only be captured incompletely. Therefore, the next logical and important step will be to develop techniques that can rapidly and, if possible, non-invasively monitor systemic immune responses. The most promising approaches involve the use of refined imaging such as magnetic resonance imaging (MRI) coupled with an appropriate 'labelling agent' that will identify antigen specific cells or other players of the immune system during the procedure. Data obtained could be fed into computers, where a whole mathematical model of the existing immune response specific to an individual can be created. This would allow us, while linked to an empirical database for pervious immune interventions and their outcome, to identify the optimal immune-based intervention for each patient and disease and monitor their therapeutic success, without having to rely on the final outcome, which is clinically undesirable.

References

Balasa B, Sarvetnick N 1996 The paradoxical effects of interleukin 10 in the immunoregulation of autoimmune diabetes. J Autoimmun 9:283–286

Bot A, Holz A, Christen U et al 2000 Local IL-4 expression in the lung reduces pulmonary influenza-virus-specific secondary cytotoxic T cell responses. Virology 269:66–77

Christen U, Wolfe T, Hughes A, Green A, Flavell R, von Herrath MG 2001 A dual effect of TNFα in type 1 diabetes. J Immnol, in press

Coon B, An L-L, Whitton JL, von Herrath MG 1999 DNA immunization to prevent autoimmune diabetes. J Clin Invest 104:189–194

Cope AP, Ettinger R, McDevitt H 1997a The role of TNF alpha and related cytokines in the development and function of the autoreactive T-cell repertoire. Res Immunol 148:307–312

Cope AP, Liblau RS, Yang XD et al 1997b Chronic tumor necrosis factor alters T cell responses by attenuating T cell receptor signaling. J Exp Med 185:1573–1584

Homann D, Holz A, Bot A et al 1999 Autoreactive CD4+ lymphocytes protect from autoimmune diabetes via bystander suppression using the IL-4/Stat6 pathway. Immunity 11:463–472

Horwitz MS, Krahl T, Fine C, Lee J, Sarvetnick N 1999 Protection from lethal coxsackievirus-induced pancreatitis by expression of gamma interferon. J Virol 73:1756–1766

Itoh M, Takahashi T, Sakaguchi N et al 1999 Thymus and autoimmunity: production of CD25+CD4+ naturally anergic and suppressive T cells as a key function of the thymus in maintaining immunologic self-tolerance. J Immunol 162:5317–5326

King C, Davies J, Mueller R et al 1998 TGF-beta1 alters APC preference, polarizing islet antigen responses toward a Th2 phenotype. Immunity 5:601–603

Klavinskis L, Whitton JL, Joly E, Oldstone MBA 1990 Vaccination and protection from a lethal viral infection: identification, incorporation, and use of a cytotoxic T lymphocyte glycoprotein epitope. Virology 178:393–400

Lee MS, Wogensen L, Shizuru J, Oldstone MBA, Sarvetnick N 1994 Pancreatic islet production of murine interleukin-10 does not inhibit immune-mediated tissue destruction. J Clin Invest 93:1332–1338

Lee MS, von Herrath MG, Reiser H, Oldstone MBA, Sarvetnick N 1995 Sensitization to self (virus) antigen by in situ expression of interferon-γ. J Clin Invest 95:486–492

Liblau RS, Singer SM, McDevitt H 1995 Th1 and Th2 CD4+ T cells in the pathogenesis of organ-specific autoimmune diseases. Immunol Today 16:34–38

Matsue H, Matsue K, Walters M, Okumura K, Yagita H, Takashima A 1999 Induction of antigen-specific immunosuppression by CD95L cDNA-transfected 'killer' dendritic cells. Nat Med 5:930–937

Racke MK, Bonomo A, Scott DE et al 1994 Cytokine-induced immune deviation as a therapy for inflammatory autoimmune disease. J Exp Med 180:1961–1966

Rocken M, Racke M, Shevach EM 1996 IL-4-induced immune deviation as antigen-specific therapy for inflammatory autoimmune disease. Immunol Today 17:225–231

Seddon B, Mason D 1999 Peripheral autoantigen induces regulatory T cells that prevent autoimmunity. J Exp Med 189:877–882

von Herrath MG 1998 Selective immunotherapy of IDDM: a discussion based on new findings from the RIP-LCMV model for autoimmune diabetes. Transplant Proc 30:4115–4121

von Herrath MG, Oldstone MBA 1997 Interferon-gamma? is essential for beta cells and development of insulin-dependent diabetes mellitus. J Exp Med 185:531–539

von Herrath MG, Guerder S, Lewicki H, Flavell R, Oldstone MBA 1995a Coexpression of B7-1 and viral ('self') transgenes in pancreatic beta-cells can break peripheral ignorance and lead to spontaneous autoimmune diabetes. Immunity 3:727–738 von Herrath MG, Allison J, Miller JF, Oldstone MBA 1995b Focal expression of interleukin-2 does not break unresponsiveness to 'self' (viral) antigen expressed in beta cells but enhances development of autoimmune disease (diabetes) after initiation of an anti-self immune response. J Clin Invest 95:477–485

von Herrath MG, Dyrberg T, Oldstone MBA 1996 Oral insulin treatment suppresses virus-induced antigen-specific destruction of beta cells and prevents autoimmune diabetes in transgenic mice. J Clin Invest 98:1324–1331

von Herrath MG, Coon B, Lewicki H, Mazarguil H, Gairin JE, Oldstone MBA 1998 *In vivo* treatment with a MHC class I-restricted blocking peptide can prevent virus-induced autoimmune diabetes. J Immunol 161:5087–5096

von Herrath MG, Coon B, Homann D, Wolfe T, Guidotti LG 1999 Thymic tolerance to only one viral protein reduces lymphocytic choriomeningitis virus-induced immunopathology and increases survival in perforin-deficient mice. J Virol 73:5918–5925

Weiner HL 1997 Oral tolerance for the treatment of autoimmune diseases. Annu Rev Med 48:341–351

Wogensen L, Huang X, Sarvetnick N 1993 Leukocyte extravasation into the pancreatic tissue in transgenic mice expressing interleukin 10 in the islets of Langerhans. J Exp Med 178:175–185

DISCUSSION

Segel: Relevant here is some work I did with my colleague, Irun Cohen, at the Weizmann Institute (Segel et al 1995). This work concerns the situation you outlined where there are aggressive cells and regulator cells. We examined this situation in the context of vaccination against autoimmune disease. Experiments by Cohen and his colleagues showed that if you give animals a certain amount of 'bad guy' autoaggressive cells, the animals get autoimmune disease. If you give fewer 'bad guy' cells, they don't develop disease. Moreover, if you follow this experiment with another experiment somewhat later, giving the standard disease-generating dose of aggressive cells, the animals still don't get autoimmune disease. Thus a lowish dose of the very same aggressive cells gives what looks like a vaccinated state.

We strove to construct the simplest possible model for these experiments with a schematic dynamic interaction between aggressive and regulator populations. In mathematical terms, the model consisted of two ordinary differential equations. As shown in Fig. 1 (*Segel*) we generated a situation with three stable states: one with a low amount of autoaggression, which we call the normal state; a second with an intermediate amount, which we call the vaccinated state; and a third with a lot of autoaggression, which we call the diseased state. Since there are three possible stable states of this dynamical system, there must be some sort of line (called 'separatrices') that will separate the possibilities. If you start on one side of the (dashed) separatrix between the first two stable states, you go to the normal state (curve A); if you start on the other side you go to the vaccinated state (curve B). There is a similar line separating the vaccinated state from the diseased state. It could be, as drawn in the figure, that this second line bends down as it moves to

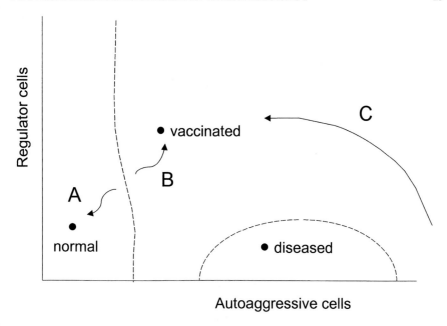

FIG. 1.   (*Segel*) Schematic showing the dynamic interaction between aggressive and regulator populations of cells in the immune system (see text for explanation).

the right. Then if you are in the diseased state and add some aggressive cells, you would bring the system into the vaccinated state (curve C). Adding more aggression can result in a less severe disease! My colleague and I had experiments in a drawer which showed exactly this. The reason is that the aggression brings forth regulation. The modelling simultaneously brings good and bad news. It is good news because our model shows conceptually how autoimmune 'vaccination' can happen. It is bad news because actual interactions are doubtless many dimensional, not just two dimensional, and it is very hard to know what is the appropriate intervention that will result in an improved outcome. For this we need precise and careful models.

Sejnowski: You seem to imply that this intermediate or vaccinated state would be stable for many years and then eventually the full blown clinical disease will develop if there is the right stimulus. It sounds a bit like it is not really a stable state, but instead a metastable state.

Segel: That sort of thing can happen. In the simplest possible model, you take certain things as constants. In fact, they aren't constants; they slowly vary. And if you slowly vary things, all of a sudden the domains of attraction may switch, and you can fall from a normal state into a bad state.

von Herrath: It is interesting how you point out that in situations of disease, if you add more aggressive cells or enhance inflammation, this may in some circumstances move the system to a vaccinated or protected state. There is now a fair amount of experimental evidence from animal models, such as those of diabetes, where this is seen (Singh 2000, Mor & Cohen 1995).

Sejnowski: Has this also been seen in humans?

von Herrath: No. In humans the real problem with autoimmune diabetes is that we don't have an effective and feasible way of collecting data for this type of disease. We can measure values in the blood, for example of antibodies, which is done very well, or oral glucose tolerance, which gives an idea of β cell function, but this is about it. Assessment of cellular autoimmune responses in the peripheral blood mononuclear cells (PBMCs) of the blood has so far been unreliable. We can't go into the pancreas, because this may cause cysts, which we don't want to risk in healthy individuals. We can't even access the pancreatic draining lymph nodes. From animal models we know that a lot of the autoimmunity happens as a cross-current between the islets and the draining nodes. This is why one has to explore the area of *in vivo* imaging systems. We would like to able to label certain cells and have a high-resolving magnetic resonance imaging (MRI) scan with which to track these autoreactive cells to the islets in real time. This would let us know where they go, how they compartmentalize, and what they go on to do. These data could then be fed into a computer analysis and give us a much better idea of how the system works.

Sejnowski: Which labels do you have in mind?

von Herrath: We are working with a group who have been tagging β cells with certain molecules which they can then visualize. The problem at this point is still the resolution of the MRI. Unless we can get it down to the single cell level, this approach will be unsuccessful.

Iyengar: You have been talking about not being able to predict. I have been talking to some engineers who do this sort of model design for a living. They have their complexity divided into what they call 'real' and 'apparent' complexity. The apparent complexity exists where they don't understand the design parameters and not because the system intrinsically behaves in a complex way. Do you think that if you could model this at a cellular level — because after all viral infection is going to be cellular — rather than at any of these higher levels of modelling, will the models give you predictive capabilities? For instance, when you talk about a second rapid infection causing infectivity to fail and your cell is saved, my challenge to you would be that unless you can show why it failed at a cellular level, doing it in islets or aggregate cells will tell you very little in terms of being able to predict outcome.

von Herrath: To model like this, you need to understand both what the virus does to the cell, and then also the cell–cell interactions. The second level of the model needs to include an organ-wide understanding of the process.

Iyengar: I would agree that you would learn something at the cellular level, but to get at infection as a whole you need a second level of model going beyond the cellular detail.

von Herrath: For example, this is how such a model could work. You start with the co-stimulators at the cellular level. If you use sophisticated imaging techniques and visualize T cell receptor clustering upon activation along with accessory molecules, one can localize the co-stimulators B7.1 and B7.2, for example, just as Mark Davis is doing (Wülfing et al 1999). One could quantitate this and get a good idea of movement within a cell. You could take these data and use them as a basis to make the cellular model. From there you can take the model to the systemic level if sufficient information is available on the trafficking of autoreactive lymphocytes (e.g. Merica et al 2000).

Sejnowski: You alluded to memory processes, which presumably take place over much longer time scales, of years.

von Herrath: Immunological memory, as well as 'autoimmune' memory, is a difficult issue: there has been a long-running controversy over whether this is maintained by persistent antigen or not. The Rolf Zinkernagel 'camp' thinks that functional immune memory is driven by antigen (Zinkernagel 2000); on the other hand, Rafi Ahmed and Polly Matzinger think it is not driven by antigen (Matzinger 1994, Whitmire et al 2000). This situation will not be easily resolved. The antigen may persist in some kind of vesicle where it is not easily detectable or 'stainable'. How a memory lymphocyte is characterized is also controversial. The markers that are used are just empirical molecules and might have nothing to do with the memory property. Most recently, it has become clear that there is some sort of homeostatic cycling of the immune system. The memory cells, although they can be long lived, turn over. The question is, how do they maintain their specificity when they are being turned over in this way (Antia et al 1998)? On the T cell level there is not a great deal of affinity maturation. It is not known how these cells turn over, and what makes them go into this maintenance cycle. It is a fascinating problem. Therefore the role of the immune/autoimmune memory is not well understood.

Kahn: I want to ask a question that might compare the first two papers that we have heard. In one situation you are talking about things which are continuous. However, when you talk about the induction of disease state, it is stochastic: you either develop diabetes or you do not. The difference between the two outcomes might be 85% killing of the β cells versus 95% killing of the β cells. At some point there will be enough cells damaged to cause a difference. The question I was trying to envision as one sets up models is this: is the power of the model decreased or increased when you are dealing with a continuous variable (such as a signalling system) versus one that is discontinuous (a stochastic event such as a disease state or mitosis)?

von Herrath: By measuring insulin levels, β cell mass and blood glucose one has a pretty good continuous variable in most experimental *in vivo* systems for diabetes (von Herrath et al 1994, Homann et al 1999).

Sejnowksi: Diagnosis of a disease state is often binary, but this hides the fact that there is usually a grey area.

Kahn: I understand diabetes very well; that is not the problem. What I am asking is, is the model less powerful because we are not able to measure the correct quantitative data or that the critical variable is not assessed at all? Or will the modelling be just as powerful if the final endpoint is the presence of absence of disease?

Iyengar: You need the trigger. I could bring my model down to this level if I didn't know about MAP kinase phosphatase. Assume that you didn't know that MAP kinase phosphatase existed. Then at certain times you put in epidermal growth factor (EGF) and the cell starts to divide. There are other cells that you put EGF into and nothing happens. This comes back to the issue of apparent complexity: because we didn't know there was this determinant process, which is the regulating enzyme, we had no idea why these cells responded to the same signal in different ways. In disease states, I suspect that there is a trigger that causes the transition. My question is whether in disease systems this trigger will be a molecular one (a single component of one cell type), or whether it involves several components from several cell types.

Sejnowski: Are you saying that if we know the initial conditions — what was there to begin with — you could predict whether an individual cell would go up or down?

Iyengar: One thing that came out of our modelling relates to the question of how memory is sustained. Our MAP kinase model gives a sustained stimulus. We don't need to preserve any individual molecule of MAP kinase. Each one can turn over and a new one can be synthesized. PKC comes back and goes through Raf, and picks up any MAP kinase that is there, so we can get continuous turnover at one end and still maintain an active state.

Dolmetsch: A philosophical question. You suggested that there are no stochastic events, only a lack of knowledge of the mechanism that underlies events. Do you think that this is the case? Do you think that if you were to know all the molecular players, everything that we now call stochastic would turn out not to be stochastic after all?

Iyengar: I wouldn't say that. I was trying to make my life easy by going along with the currently favoured idea of molecular scaffolds and anchors, which in a computational sense makes our life a lot easier. Clearly, there are many processes that are stochastic. There are probably real stochastic processes and real uncertainty, which means that however much we know, there will probably be some variability in our prediction. Until we actually measure everything and prove that it is there, we can't say that it is. At this stage, I would still use the engineers' concept of apparent complexity where we haven't measured everything correctly.

Sejnowski: There are known sources of fluctuations. Diffusion is clearly an important process: we have to live with the variability with which a single molecule will diffuse from point A to point B. As an example, let us take the simulation of the release of acetylcholine and its binding. We can start from exactly the same initial positions, use a random walk model, and see the same randomly fluctuating currents that are seen physiologically at the neuromuscular junction. We have to do this computation dozens of times and average, just as the physiologist does, to get a good result. This is an inherent source of stochasticity, which nature can take advantage of as a computational principle to overcome barriers.

Dolmetsch: It is analogous to the difference between thermodynamics and statistical mechanics. If you have lots of molecules, you can predict what they are going to do, but it is much harder to do this with just a few. In the disease state, there is prediabetes for a period of, say, seven years, then one day you develop diabetes. Is this truly predictable? Is it that we don't know some variable, and if this variable were known then prediction would be possible? It might be that this is not the case, and that one day, one cell does something for some reason, and this somehow nucleates the disease. It might be very difficult to predict. A better example of this is probably cancer, in which there are a certain number of hits, which are stochastic.

Iyengar: I think we can predict cancer pretty well.

Sejnowski: It is probably the case that there are some things which we can make a definite prediction about, and other things about which you can only make a probabilistic prediction. The question is perhaps a philosophical one, ultimately.

Iyengar: That is a multicellular question. But cancer is basically a unicellular disease. Only one cell needs to transform and then it takes over.

Brenner: We don't know that. There may be many such initial events, and they just decay. There may have to be some other stochastic condition that nucleates the disease. We know quite a lot about these things in ecological systems and it may be helpful to apply ecological 'population biology' thinking when we consider populations of cells in a complex environment.

References

Antia R, Pilyugin SS, Ahmed R 1998 Models of immune memory: on the role of cross-reactive stimulation, competition, and homeostasis in maintaining immune memory. Proc Natl Acad Sci USA 95:14926–14931

Homann D, Holz A, Bot A et al 1999 CD4+ T cells protect from autoimmune diabetes via bystander suppression using the IL-4/Stat6 pathway. Immunity 11:463–472

Matzinger P 1994 Immunology. Memories are made of this? Nature 369:605–606

Merica R, Khoruts A, Pape KA, Reinhardt RL, Jenkins MK 2000 Antigen-experienced CD4 T cells display a reduced capacity for clonal expansion in vivo that is imposed by factors present in the immune host. J Immunol 164:4551–4557

Mor F, Cohen IR 1995 Vaccines to prevent and treat autoimmune diseases. Int Arch Allergy Immunol 108:345–349

Segel LA, Jaeger E, Elias D, Cohen IR 1995 A quantitative model of autoimmune disease and T-cell vaccination: why more cells may produce less effect. Immunol Today 16:80–84

Singh B 2000 Stimulation of the developing immune system can prevent autoimmunity. J Autoimmun 14:15–22

von Herrath MG, Dockter J, Oldstone MB 1994 How virus induces a rapid or slow onset insulin-dependent diabetes mellitus in a transgenic model. Immunity 1:231–242

Whitmire JK, Murali-Krishna K, Altman J, Ahmed R 2000 Antiviral CD4 and CD8 T-cell memory: differences in the size of the response and activation requirements. Philos Trans R Soc Lond B Biol Sci 355:373–379

Wülfing C, Chien YH, Davis MM 1999 Visualizing lymphocyte recognition. Immunol Cell Biol 77:186–187

Zinkernagel R 2000 On immunological memory. Philos Trans R Soc Lond B Biol Sci 355:369–371

Controlling the immune system: diffuse feedback via a diffuse informational network

Lee A. Segel

Department of Computer Science and Applied Mathematics, The Weizmann Institute of Science, Rehovot 76100, Israel

Abstract. Diffuse feedback is defined to be a process by which a system in some sense improves its performance with respect to a variety of conflicting and even contradictory goals. In the immune system, such feedback is mediated by scores of extracellular chemicals (cytokines), each of which participates in achieving several goals. Progress toward any given goal is mediated by several cytokines. The 'immunoinformatics' of this diffuse informational network will be discussed. It will be shown how diffuse feedbacks, based on this network, can improve the performance of a given type of immune effector cell, and can cause the preferential amplification of more potent effectors. It will be argued that diffuse feedback also acts in other biological systems ranging from the metabolic system to ant colonies.

2001 Complexity in biological information processing. Wiley, Chichester (Novartis Foundation Symposium 239) p 31–44

The immune system is a superb venue for learning about biological information processing. Because of the immune system's intrinsic interest and medical importance, its 'hardware' is rather well understood, although much remains to be done. At its molecular level the remarkable phenomenon of hypermutation chemically scrambles genetic information in order to provide diversity for B cell receptors. But what interests me more is the cellular level — because I believe that insights at this level are not only definitive with regard to immune system behaviour, but also are applicable to other major biological systems, and indeed to non-biological distributed autonomous systems.

Vertebrates possess trillions of immune cells, of dozens of different types, with no apparent 'boss'. Different sets of cell types are mobilized to combat different species and strains of pathogens that attack the host. Moreover, the immune system participates in other homeostatic tasks such as wound healing and tissue remodelling. Scores of signalling molecules, called cytokines, guide the immune system. Each cytokine seems to have several functions, and any given function seems to be affected by several cytokines. When suitable receptors are ligated, not one but several cytokines are typically secreted.

How does this vastly complicated distributed autonomous system 'decide' what to do and when and how intensely to do it? I will discuss various aspects of this question, emphasizing the role of information. In particular I argue that a decisive role is played by what I call a *diffuse informational network*, based on cytokines. In doing so I am responding to the suggestion of Orosz (2001) concerning the key role of 'immunoinformatics', defined to be the study of 'how the immune system generates, posts, processes, and stores information'.

I will not give references to well-accepted assertions about immune system operation. The reader who wishes to learn more can consult texts such as that of Janeway & Travers (1997) or that of Paul (1999). I have concentrated here on showing that my ideas for the role of information in immunology fulfil a need and are feasible. In addition, evidence is required that these ideas are actually implemented. For that, see Segel & Bar-Or (1999).

Cytokines: command network or informational network?

The immune system is triggered to act by information that something is wrong. The following are non-exclusive alternatives for the triggering mechanism.

  (i) Characteristic microbial molecules bind special 'pattern recognition detectors'. These are receptors on cells, such as macrophages, of the evolutionarily primitive innate immune system (Janeway 1992).
 (ii) A 'tuneable activation threshold' detects significant departures from 'normal' conditions (Grossman & Paul 1992).
(iii) Special receptors on various cells bind molecules that signal some form of 'danger' or tissue destruction (Matzinger 1994, Ibrahim et al 1995).

Once triggered, the immune system's response is normally regarded as *reactive*. A number of factors combine to shape the response — not only the initial pattern detectors but also receptors that detect peculiar molecular constituents of the individual antigen. Also of importance are the different conditions that are characteristic of the various tissues. All these factors interact in a complex manner to yield a response that has been selected by evolution to be advantageous to the host.

How do the cytokines modulate the immune response? The classical view is that the cytokines form a *command network* that directs cell activity. For example, *in vitro* experiments show that the switch of B cells from secreting IgM antibody to the alternative IgG can be induced by interleukin (IL)-2, IL-4, IL-6 and interferon

(INF)-γ. I advocate the (non-exclusive) alternative that the cytokines can be regarded as forming an *informational network*. Typically by 'chords' of several cytokines, not by 'tones' (single cytokines), this network provides information that various different cell types can use in different ways to forward the goals of the immune system. In fact 'hymns' of cytokines is a better metaphor, because it is likely that there is significance to the temporal development of cytokine profiles.

The 'command' view of cytokines is reflected in numerous surveys of cytokine action, which typically list the various activities associated with a given cytokine, as well as the receptor for that cytokine (for example, see Appendix II of Janeway & Travers 1997). Viewing the cytokines as an informational network focuses attention differently, on what molecules trigger (via what receptors) secretion of the various cytokines, in what amount. For example, since we know that engagement of the lipopolysaccharide (LPS) receptor on macrophages induces secretion of IL-1, IL-6, IL-8, IL-12 and tumour necrosis factor (TNF)α, then we can infer that this cytokine chord gives the information 'there is potential danger from Gram-negative bacteria' (whose outer membrane contains much LPS). Unfortunately, information is not readily available on the amplitudes of the various secretions, for presumably several different 'chords' are composed of the same cytokine ingredients in different proportions.

Another of the many macrophage receptors (an Fc-γ receptor) binds the constant 'γ region' of IgG, typically during the processes of macrophage phagocytosis of pathogens that have been 'opsonized' by binding IgG antibodies of suitable specificity. IL-10 secretion is induced by such binding, so that a message associated with abundant quantities of this cytokine is 'macrophages are internalizing opsonized pathogens'. (I am using 'message' in its general sense, not in connection with messenger RNA.) In addition the message 'apoptosis is occurring' can be inferred by relatively high levels of transforming growth factor (TGF)β, PGE2 and PAF, which (*in vitro*) are secreted by macrophages that have ingested apoptotic cells (Fadok et al 1998). Such secretions when, respectively, Fc-γ receptors or apoptotic body receptors are bound constitute examples of what I have termed the 'do-moo' principle — if a cell is accomplishing one of the actions that it is capable of then it 'moos' — it informs the whole system of what it is doing.

Once an information network has been postulated, a natural question is whether various cells respond appropriately to the available information. For example, if macrophages can effectively dispose of a pathogen challenge by ingesting opsonized pathogens then one would expect to see a down-regulation of alternative responses via T cells and an up-regulation of antibody-secreting B cells. (This illustrates an important principle — different cells should respond differently to the same information.) In broad terms, this expectation is confirmed by data reviewed by Denny (2001). Increasing IL-10 curbs antigen presentation by macrophages and monocytes (their blood-borne precursors), which decreases T cell activity. IL-10 also acts to encourage B cell proliferation and differentiation. Moreover, at least in monocytes, there is positive feedback from this type of IL-10 secretion, since the secretant up-regulates the Fc-γ receptors whose ligation induces the secretion in question.

The reader should not gain the impression from the previous paragraph that adopting an information-network point of view provides instant clarification of the function of cytokine networks. At best, it is a promising lead in understanding cytokines (concerning which more than 10 000 papers a year are currently being written, yet 'practically nothing is known about the behaviour of the network as a whole'; Callard et al 1999). Thus, if we return to the example of IL-10, we must bear in mind that it is not only ligation of Fc-γ receptors on macrophages that induce its secretion. Among several other possibilities is infection of B-cells by Epstein–Barr virus and the appearance of certain types of lymphomas (Denny 2001). Thus there appears to be no single 'message' that can be attributed to IL-10. Moreover, there are other actions of IL-10 in addition to those I have cited. For example the up-regulation of Fc-γ receptors is associated with additional antibody-dependent cell killing by natural killer (NK) cells (Denny 2001). This action seems to have little to do with the macrophage ingestion of opsonized pathogens that was previously discussed.

To provide a tentative explanation for the complications just cited, I suggest that the system can cope with a message of the form 'either I or II is happening', for this narrows the focus of the system. Mobilizing NK cells might well be an appropriate response to 'alternative II' for interpreting high IL-10 concentration, the presence of lymphomas. If responses of the immune system are generally encouraged when they lead to the proven repelling of dangerous attackers (see below) then there would be an eventual selection of the correct response to the ambiguous message 'it's I' or 'it's II': suitable antibody secretion and opsonization if the threat is bacteria that can be destroyed by macrophages via opsonization, and active NK killing if the threat is a lymphoma.

Another complication inherent in the view of cytokines as information providers is the ability even of mere viruses to subvert the signalling system (Ahmed & Biron 1999). Here are three examples.

(i)   Pox viruses encode a protein, T2, that is similar to the TNF receptor. T2 is released from infected cells; it binds TNF, presumably blocking TNF's antiviral action.

(ii) A protein encoded by the Epstein–Barr virus blocks the synthesis of IL-2 and INF-γ.

(iii) A protein encoded by the myxoma virus binds INF-γ.

An argument frequently made against the notion that an information network plays an important role in immune defence is that pathogens can subvert the information. As I have just pointed out, subversion is indeed possible. Yet, somehow the subverting pathogens are not devastating, which indicates that the subversion does not force the host to abandon all use of the information in question. Rather, an alternative is found. Indeed, the diffuse nature of the cytokine informational network may well be one defence against subversion.

How diffuse feedback can modify the original immune response

I do not accept the conventional view that the immune response is essentially pre-programmed, even though features such as those that I have discussed can provide a very sophisticated hard-wired response. For one thing, the immune system is so varied and complex that a hard-wired response cannot be relied on. Many aspects of the response are unpredictable (Sercarz 2001). Moreover the attacking pathogens mutate so fast that they are likely to escape any inflexible attack.

The pre-programmed response is the first attempt, and may well be a successful one, but I believe that the first response is often modified, perhaps in a major way, during the period that the host is subject to a given challenge to homeostasis. Here's how I suggest that the modification operates.

(A) The dominant initial immune response is biased by evolutionary experience to provide adequate countermeasures for combating a wide variety of 'standard' pathogens and other 'standard' disturbances to homeostasis. But the initial response contains a number of other elements, in addition to the dominant response, which provide a spectrum of replies to each challenge.

(B) The immune system can be regarded as possessing a number of goals, which typically overlap or even contradict each other. A set of sensors monitors progress toward these goals.

(C) A diffuse informational network presents to all cells not only indications of progress toward the various goals but also other important information concerning the state of the immune system and of the host.

(D) During the course of any challenge, by a process of *diffuse feedback* the immune system continually adjusts its response so as to obtain some sort of overall improvement in attaining its goals. The fundamental hypothesis here is that the information mentioned in Item C is employed by individual cells to hone their characteristic response, and by the system as a whole to select for relative amplification of those cells in the response spectrum (Item A) that are most effective in promoting the overall goals (B).

Two major immune system goals can be expressed by the postulates that the immune system acts as if — all other things being equal — killing dangerous pathogens is good and harming self is bad. There is no teleology here. 'Harming self is bad' is merely a terse formulation of the observation that, during the course of evolution, selection processes generally favour organisms whose immune systems — all other things being equal — do not produce excessive self-harm.

To modify its behaviour in such a way as to improve in some sense its attainment of its multiple goals, the immune system must know 'how it's doing'. Given the goals I have suggested, this suggests the presence of a 'harm chemical' H and a kill chemical K that respectively provide evidence of harm done to the host and of pathogen killing. (See Segel & Bar-Or (1999) for specific suggestions for molecules that act as harm and kill chemicals.) (If tumour cells can also be recognized as attackers, then there should be chemicals that give evidence of tumour destruction.)

Evidence of harm by pathogens should up-regulate immune response (for then the pathogens are dangerous) but response should be down-regulated if the harm is generated by the immune system. These two types of harm, H_P and H_I, can be distinguished if they are respectively associated with high pathogen levels or high levels of immune activity. Moreover, association of H_P and K, i.e. high levels of both pathogen harm and pathogen killing, signals the achievement of the goal of destroying dangerous pathogens. By contrast, high levels of H_I are bad; they indicate that the immune system is damaging the host. Simultaneous consideration of both good and bad features of a given response is not difficult; the response intensity need only be programmed to be an increasing function of evidence of 'good' (e.g. simultaneous high values of both K and H_P) and a decreasing function of evidence of 'bad' (e.g. high values of H_I).

Segel & Bar-Or (1999) provided a mathematical model that showed how the principle just cited could improve the performance of a single cell. The basic problem that they examined is one of wide application in immunology, arranging an immune response with a suitable level of inflammation. Too little inflammation engenders a poor immune response; too much engenders unnecessary self-damage. This idea was examined in the framework of the hypothesis that the immune system has evolved with the *overall goal* of minimizing the combined direct and indirect harm that stems from a pathogen, where the indirect harm comes from inflammation. The analysis showed that if the pathogen characteristics are fixed then there is an optimal level of inflammation. But this level depends on pathogen virulence. For example, faster growing pathogens should evince a higher level of inflammation, since it is worth suffering increased damage from inflammation in order to prevent even more damage from the rapidly reproducing pathogens. It was demonstrated that a pre-programmed immune system cannot cope well with a variety of pathogens but that 'coping' can be greatly improved with the aid of feedbacks governed by information concerning concentrations of the chemicals H and K.

I stress the difference between the *overall goal* of minimizing total damage to the host by both pathogens and the immune system and what can be temporarily termed the *intermediate goals* of killing dangerous pathogens and avoiding harm to self. As we have seen, the intermediate goals overlap and conflict. No optima are sought: optimal achievement of one intermediate goal may well mean that an important alternative goal is almost entirely neglected. The degree of achievement of intermediate goals can be monitored and they moderate day to day and week to week activity of the immune system. Although immune 'goals' are an abstraction invented by scientists, the intermediate goals seem to model parts of immune system operation with considerable faithfulness.

The relative importance of the various intermediate goals is decided by evolution. By contrast an overall goal, such as minimizing total harm from pathogens and the immune system (a way of 'maximizing fitness'), is too lofty to provide a basis for the development of an effective immune response to a particular challenge. An overall goal is a theoretical construct that is aimed to underpin a computationally feasible substitute for the true complexities of evolution.

In addition to providing a means for improving the operation of a particular effector cell, information on immune performance can permit the selection of those effector cells that contribute strongly to the performance of a given immune task, and to the down-regulation of inefficient cell classes. This process was also modelled by Segel & Bar-Or (1999). A key question here is this. Suppose that informational chemicals indicate that a certain task (for example killing dangerous pathogens without causing too much self-damage) is being well done. How can the 'credit' be assigned to the right set of effectors? One way is by spatial organization. Suppose that several different effector types are concentrated at several points of space. Suppose further that a chemical C is secreted wherever the given task is well done, and suppose finally that *all* effectors are up-regulated (via proliferation and/or activation) by high concentrations of C. This is enough to ensure the selection of 'effective effectors'. If effective effectors can be selected, then feedback can help select between the helper T cell classes Th1 and Th2 (Segel 2001a).

Another way to describe the process is as follows. As I have emphasized, there is a spectrum of immune responses. The effects of each response are 'simulated' at different points of space—for example in lymph nodes or spleen, immune organs that are noted for their subtle spatial organization. Those responses that prove their worth are magnified.

Discussion: 'informatics' of other complex biological systems

In my analysis of 'immunoinformatics' I have concentrated on the suggestion that the cytokines form a diffuse informational network. But it is not possible to consider biological information in isolation from how it is used. This has required discussions about diffuse feedback, which employs the information to modulate immune activity in order to promote a set of overlapping and contradictory goals.

The approach described here offers explanations for salient observations concerning the cytokine network. Why does a given cytokine 'command' numerous functions? Because different cells respond differently to the same information. Why is a given function affected by so many cytokines? Because this function is relevant to many overlapping and contradictory goals.

The approach that I have advocated is relevant to other biological contexts (Segel 2001b). For example, the metabolic system must also contend with a variety of overlapping and contradictory requirements, as it ceaselessly adjusts itself in the face of changing nutritional opportunities and changing structural and energetic demands. Here the metabolite concentrations themselves seem directly to provide information on the state of the network. The presence of multiple regulatory sites on key enzymes (e.g. at least six on phospho-fructokinase) seems sufficient evidence for the assertion that the information is exploited via diffuse feedback.

The genomic regulatory network is another locus for diffuse feedback. Yuh et al (1998) demonstrated that the control of the sea urchin developmental gene *Endo16* is exerted via seven sites on the *cis*-regulatory element 'module A', which acts as a complex switch in processing occupancy information concerning these sites. Here too there are multiple 'goals', for *Endo16* seems to have many functions, which are exerted at different locations and at different times in the life cycle of the sea urchin. It remains to elucidate how information about the goals is translated into varied concentrations of DNA binding proteins.

An additional example of 'informatics' in biology, concerns information transfer in ant colonies. (i) As discussed in the review by Hölldobler (1995), there are documented examples of signalling 'chords': for example, mymicine worker ants respond maximally to trail pheromones composed of a 3:7 ratio of two pyrazines. (ii) Hölldobler conjectures that if genetically similar members of an ant colony tend to produce similar patterns of multicomponent alarm signals then this can inform the ants whether nest mates or aliens are producing the signal. (iii) In the immune system, different effector cells are selected. In ant colonies, adaptive behaviour is more likely to occur when ants switch tasks. The 'do-moo' principle could well hold, however, for ants from different task groups (e.g. foragers and nest maintenance workers) have different chemical profiles (Wagner et al 1998). These differences may be sensed in brief antennal contacts.

Returning to immunology, let me mention a possible medical application of the point of view that it is preferable to regard the cytokine network as providing information, not as embodying commands. The command paradigm leads to the idea of accomplishing the beneficial proliferation of cell type A by introducing a chemical α, that is known to accelerate the division of A cells. But α may have many side affects, and may not work well owing to the presence of α antagonists. It may be better to inject a panel of cytokines which the body interprets as information that it is in a situation where high A concentrations are beneficial. Then the body itself will come up with a good way to achieve such concentrations.

References

Ahmed R, Biron CA 1999 Immunity to viruses. In: Paul WE (ed) Fundamental Immunology. Lippincott-Raven, Philadelphia, p 1295–1334

Callard R, George AJT, Stark J 1999 Cytokines, chaos and complexity. Immunity 11:507–513

Denny T 2001 Cytokines — a common signalling system for cell growth, inflammation, immunity and differentiation. In: Segel LA, Cohen I (eds) Design principles for the immune system and other distributed autonomous systems. Oxford University Press, Oxford, p 29–78

Fadok VA, Bratton DL, Konowal A, Freed PW, Westcott JY, Henson PM 1998 Macrophages that have ingested apoptotic cells *in vitro* inhibit proinflammatory cytokine production through autocrine/paracrine mechanisms involving TGFβ, PGE2, and PAF. J Clin Invest 101:890–898

Grossman Z, Paul WE 1992 Adaptive cellular interactions in the immune system: the tunable activation threshold and the significance of subthreshold responses. Proc Natl Acad Sci USA 89:10365–10369

Hölldobler B 1995 The chemistry of social regulation: multicomponent signals in ant societies. Proc Natl Acad Sci USA 92:19–22

Ibrahim MA, Chain BM, Katz DR 1995 The injured cell: the role of the dendritic cell system as a sentinel receptor pathway. Immunol Today 16:181–186

Janeway CA Jr 1992 The immune system evolved to discriminate infectious nonself from noninfectious self. Immunol Today 13:11–16

Janeway CA Jr, Travers P 1997 Immunobiology. Blackwell Scientific, Boston, MA

Matzinger P 1994 Tolerance, danger, and the extended family. Ann Rev Immunol 12:991–1045

Orosz C 2001 An introduction to immuno-ecology and immuno-informatics. In: Segel LA, Cohen I (eds) Design principles for the immune system and other distributed autonomous systems. Oxford University Press, Oxford, p 125–150

Paul W 1999 Fundamental Immunology. Lippincott-Raven, Philadelphia, PA

Segel LA 2001a How can perception of context improve the immune response. In: Steinman L (ed) Autoimmunity and emerging diseases. Center for the Study of Emerging Diseases, Jerusalem, p 169–191

Segel LA 2001b Diffuse feedback from a diffuse informational network: in the immune system and other distributed autonomous systems. In: Segel LA, Cohen I (eds) Design principles for the immune system and other distributed autonomous systems. Oxford University Press, Oxford, p 203–226

Segel LA, Bar-Or RL 1999 On the role of feedback in promoting conflicting goals of the adaptive immune system. J Immunol 163:1342–1349

Sercarz E 2001 Distributed, anarchic immune organization: semi-autonomous golems at work. In: Segel LA, Cohen I (eds) Design principles for the immune system and other distributed autonomous systems. Oxford University Press, Oxford, p 241–259

Wagner DM, Brown JF, Broun P et al 1998 Task-related differences in the cuticular hydrocarbon composition of harvester ants, *Pogonomyrmex barbatus*. J Chem Ecol 24:2021–2038

Yuh CH, Bolouri H, Davidson EH 1998 Genomic *cis*-regulatory logic: experimental and
computational analysis of a sea urchin gene. Science 279:1896–1902

DISCUSSION

Dolmetsch: Things are a bit different in the case of metabolism: you have the end
product feeding back on the metabolic enzymes, but in killing bacteria what you
really need is a signal for success. This may not be information that the immune cell
itself has. It may be that this signal is something that feeds back from the bacteria.

Segel: Yes, some sort of signal for success is needed to feed back. What would be
success when you are thinking about killing bacteria? A dead bacterium. But you
don't need a corpse; it is sufficient to have a scalp — irrefutable evidence that there
is a corpse somewhere. I claim that there are indeed 'scalp' molecules (see Segel &
Bar-Or 1999). One example is mycolic acid, which is a constituent of the cell wall of
a certain class of Gram-negative bacteria — not of the outside cell wall, as in the
case of LPS, but of the inside cell wall. Mycolic acid binds to CD1, an MHC-like
molecule that 'presents' lipids and thus has a role in modulating the immune
response. Other 'scalp' molecules for bacteria are *N*-formyl peptides that
promote inflammation by attracting leukocytes.

In general, I suggest that with its reflexive responses, the built-in innate immune
system has all sorts of ways to take care of the classical kinds of pathogens that have
been around and will continue to be around, because we have coevolved nicely
with them. More 'sophisticated' immune responses with feedback on the fly
during the course of the hours and weeks that a single disease may last, these
have evolved to take care of the 'wise guys': pathogens that rapidly evolve a new
way to 'get us'.

Brenner: I disagree with your anthropomorphic view. I put this in the same genre
as Dawkins' 'selfish gene'. It may be a nice way of looking at the problem, but it is
very misleading, because it doesn't connect you to what you have to really solve.
The immune system isn't everywhere; only some animals have it. I will give you a
counter-argument to the LPS idea you proposed. Once upon a time there wasn't an
immune system. Defence mechanisms certainly evolved to cope with a generic
microorganism. One way of detecting a Gram-negative bacterium is to use a
receptor to bind LPS. Therefore this was not designed as a signal to give the
body information about how well you are doing, as you want to claim. It is just
an ancient defence mechanism which has been overlaid through the course of time
with this very elaborate immune system. There are many other examples like this.
Evolution measures reproductive success, and the systems we have today are those
that worked, while the ones that didn't have been wiped out. The complexity of the
cytokine network did not arise to provide information. It is complex because it has
been used to control the development of the immune system which needs many

cytokines because it has many different cell types. We have to consider the existence of conditions such as, 'If cytokine x and y but not z is present', then the receptors will signal this perhaps to turn on a gene which allows that cell to move to the next stage. People like to imagine that there are goals in biological systems, but I have always thought that once you go down that path, the thinking you apply then becomes perverted. We also need to ask where the molecular information comes from. Receptors cannot arise from nothing, and all have origins further back. They may have been used for different things; our job is to show how they got harnessed into this developmental pathway, which is what the immune system is. It is not a question of taste, but one of discipline.

Sejnowski: Is it known whether it is a Boolean operation?

Brenner: I think it is a Boolean operation plus quantitative thresholds. It is necessary to get away from the idea of commands — these are not instructing a cell what to do next, but rather are providing the conditions that permit the cell to go on to the next stage, and withdraw the next book from the library of genes and read out what it has to do. It is more a library model than an instructive model.

Segel: I would answer with the following. First, the instructive view should be modified to something more complicated, a program view: if A and not too much B, then C in an amount depending on D and E. The program view is more accurate, but it is still a model. And of course the immune system doesn't have anything 'in mind': it behaves, it does, it was selected. None the less, my multiple goals model, taken carefully, can be a useful aid to understanding. Physiologists certainly talk about the homeostatic system that keeps our body at a particular temperature and keeps the concentration of CO_2, cholesterol, glucose and so on within appropriate ranges. In essence, they postulate 'goals' for homeostasis, and everyone thinks that this is a useful way of working. If you just take that well accepted physiological view one step further, and allow for the fact that these different, well accepted model goals interfere and overlap, then you are with me. It is not so radical.

Brenner: Let me make one other point, concerning phosphofructokinase. You say it is more complicated, but if you think about it, it is a device to measure ratios. This may be what you want.

Segel: Why does it want to measure so many different ratios? Because it is doing several different things at the same time. If you read Stryer (1988), he talks about several different functions that phosphofructokinase is simultaneously involved in.

Brenner: This enzyme has partitioned functions. It isn't sitting there in schizophrenic doubt, not knowing what to do next.

Sejnowski: If I could reformulate this issue, I think what Lee Segel is saying is that when a cytokine signal increases it is giving the rest of the system information that some event is occurring, but it is actually a well defined piece of information, as opposed to this binary or Boolean model.

Brenner: I am saying that there is a contrast between those two views. We don't know which is the right one. The rate of an enzyme reaction can certainly depend on more than one parameter, and a metabolic system can equally be optimized for more than one signal.

Iyengar: I didn't understand one point about the organization of information. When you have these information sub-goals, when you classify these, does this aggregate define your system? This comes to your question of what each part does in respect to the other. If there are three cytokines, for example, do they show up together or does one show up just late enough so that it can or cannot have an effect? How do you define your information sub-goals in the way you model these systems?

Segel: One of the things that as yet I don't understand very well is how I can take an abstract system and define what goals or sub-goals it can usefully be regarded as having. We all agree that an organism works to enhance its own reproduction in the long run. But this doesn't help a given cell to decide what to do in a particular circumstance. The cell needs proximal information from many different sensors in order to tell it how to behave at a given moment. My model of multiple goals, if successful, is still no more than a model. This means that if I can think of several suitable goals, then perhaps I can explain 75% of a certain amount of information. As with every model, there will be things I can't explain and questions that I can't touch.

Dolmetsch: Let's approach this from an experimental point of view. The idea of feedback is well established in biology, and clearly there is feedback at many levels and from multiple systems. In a model like yours, is there any way in which you can predict mathematically what I want to look for experimentally as a feedback mechanism? If I was to define the initial state and the final state, could you make a prediction as to the minimum number of feedbacks there are? Could you give me a concrete idea of what sort of experiment I should do?

Segel: Here is one possibility, dressing a sheep in wolf's clothing. Most immunologists believe that it is the epitopes on a pathogen that define the response. Take some standard Gram-positive bacterium and stick a whole bunch of LPS on the outside, so that the bacterium has Gram-negative clothes. Then, see whether after the first response due to the ligation of the LPS receptor there is a shift in response when the system figures out that the standard thing that it is doing isn't working out. Another experimental guideline is a shift in emphasis. In addition to determining many factors that bias choice towards Th1 or Th2, examine how factors lead to the correct choice in the Th1/Th2 dichotomy.

Laughlin: You repeatedly used the word 'information'. An information theorist would say that in this system you should use combinations of different cytokines to specify a set of probabilities, which would specify the likelihoods that the cell did each of the several different things that it was able to do. Have you applied that approach? It gets over the problem of focusing diffuse signals onto a single action.

Segel: You are right that it would be an improvement in my modelling (though a complication) to use information to regulate probabilities. Concerning signal combinations, I have suggested specific chemicals which indicate that the body is being harmed. It makes a big difference whether the harm is due to the immune system (all other things being equal, this should down-regulate the immune system) or the harm is being done by pathogens (which should up-regulate the system). How could you tell the difference between harm done by the immune system and harm done by the pathogen? If harm is associated with a high level of pathogens, then the best bet is that it is being done by pathogens. If harm is not associated with high levels of pathogens, the best bet is that the harm is being done by the immune system. Here is a way to associate two pieces of information to give you additional information.

Laughlin: Information theory would ultimately tell you how many signals are required to do this.

Segel: A graduate student has started working on a project involving a genetic algorithm as a surrogate for evolution. Suppose we put in a panoply of possible signals, with parameters giving the strength of those signals. How will the system evolve so that the signals are used to minimize the overall damage from pathogens? Are there circumstances wherein the system will use information in the way we suggested? This is a theoretical test of the ideas that I have proposed.

Iyengar: What would an endpoint measure be in your model?

Segel: Now you are getting into the real nitty gritty details. I don't think we can say at the moment. As an example of what might be done (see Bergman et al 2001), we could take a restricted system and try to make sense of the Th1/Th2 choice. There is evidence that the immune system tends to shift to Th2 at high pathogen levels, and that initially there is a Th1 bias. If so, our modelling shows that the Th1/Th2 choice is usually appropriate, except that fast-growing Th1 pathogens are a problem. In order to handle this case, outside information is required. This is a highly complex system that we have modelled in a simple way. We have lumped Th1 cytokines and Th2 cytokines together, instead of all the dozens of different ones, and we have incorporated some information. I don't think we will be able to say something like, 'We predict that if you measure TGFα its concentration will triple after 4 hours, otherwise we will jump off a bridge', for some years yet.

Brenner: There are now many examples that we can apply to this question. A group of people, who die of cerebral malaria in Africa, have been shown to harbour a mutation in the promoter of the NF-κB recognition site of the TNF gene. Whatever signal these people are getting from the malaria, they are unable to respond by making adequate TNF. When we test wild-type mice with viruses and bacteria, we find many strains differing in their resistance. The genes can then be examined to provide information on what evolution has done to optimize survival after infection with different pathogens. The genetic approach will also uncover mechanisms of resistance. I am sure there will be many surprises.

Segel: We have this complex cytokine network and we don't really understand it, but we want to cure sick people. Immunology is a funny subject: if you don't start curing people within 10 years, people think that you are not a good immunologist. The command method doesn't work very well: if we add a particular interleukin, it often turns out there are all sorts of side effects and it doesn't do what we want. For example, we might want to cure someone by switching from Th2 to Th1, so we add something which is supposed to push the system in that way. If this 'command mode' doesn't work, we can try an alternative strategy of tricking the system by providing it with some sort of information via a panel of cytokines, information that implies that the right sort of response is to switch to Th1. Then we can let the different cells themselves differently interpret this information so that the whole system switches to Th1. It is an alternative philosophy that may prove useful.

Brenner: That's voodoo.

References

Bergmann C, van Hemmen JL, Segel LA 2001 Th1 or Th2: How an appropriate T helper response can be made. Bull Math Biol 63:405–430

Segel LA, Bar-Or RL 1999 On the role of feedback in promoting conflicting goals of the adaptive immune system. J Immunol 163:1342–1349

Stryer L 1988 Biochemistry, 3rd edn. WH Freeman, New York

General discussion I

Sejnowski: I have a list of issues that we might address in this general discussion, prompted by the first three papers. The first had to do with this issue of stochastic variability and how this should be dealt with mathematically. Sydney Brenner, during the break you came up with a nice experiment that can be used to examine a particular probabilistic approach: could you summarize this?

Brenner: This is an experiment carried out by Novick and Weiner more than 40 years ago. In *Escherichia coli*, the Lac operon specifies a β-galactosidase and a permease, and can be induced with isopropylthiogalactoside (IPTG). The permease is needed for the inducer to get into the cell. However, there is another leaky permease for IPTG, so at a very high concentration (10^{-3} M), it can get in. You now take a population of uninduced bacteria and put them in high IPTG (10^{-3} M) for various periods, and then dilute them into a medium containing the maintenance level of IPTG, which is 10^{-6} M. Those that have had enough permease induced during the initial period now become fully induced and loaded with β galactosidase, and this state is passed on to their progeny. Those that do not have enough permease stay uninduced. The culture therefore differentiates into two kinds of cells, fully induced and uninduced. The ratios are maintained, except that the ones with lots of β-galactosidase grow a bit more slowly because they are at a disadvantage. It is clear that the initial induction is a stochastic event. There is of the order of 100 molecules of repressor in each cell, which is a femtolitre in volume. There are also cells that apparently lose the property of being fully induced. This can be explained by partitioning of the relatively small number of permease molecules at each division. Presumably, instances arise where one daughter cell receives a number of permeases that is below threshold, and in subsequent divisions the molecules will then dilute out and cells that have no enzyme will ultimately appear. This is a very good model system: you can do experiments on this and under certain conditions you can make fairly accurate predictions, relating the numbers of cells induced as a function of the initial induction time. But if you want to describe the process at a quantitative level, at the level of causal relations involving molecules, first you need to understand the details of the process and use a stochastic treatment.

Sejnowski: This is similar to the mathematics one sees in population biology and epidemiology for keeping track of probability distributions in populations.

Brenner: That is right.

Iyengar: What happens when you have a pool of molecules, such as with MAP kinase, when they go back and forth between a probabilistic set of events and a deterministic set? This is like when MAP kinase climbs on and off these scaffolds. Once it is on the scaffold it can nicely be phosphorylated. What kind of quantitative representations could be used here? There are probably many molecules that behave this way within the cell. They are concentrated in a certain region, then there is a gradient, and at the lower end of the gradient they behave in a stochastic fashion.

Sejnowski: You can imagine that it is just a matter of numbers. If the numbers drop below a certain value, fluctuations become as large as the numbers.

Brenner: With many of these processes, we can place limits to the variation in number and how much can be tolerated. In general, an organism mutant in one of the copies in genes specifying these proteins has only half the normal amount of protein and function, but is still normal. That is, the mutation is recessive in the heterozygote. Because function is normal, we know that variation of protein concentration by a factor of two can be tolerated. People have not used these facts as constraints for modelling. If half works as well, this says quite a lot about what the system can and can't do. There are other constraints that we can find. For example, in tetraploid cells, such as Purkinje cells in the cerebellum, there is twice as much DNA and the cell will have twice the volume. More protein will be produced from the added gene copies but of course there will be different consequences for those that are in solution and those that are in a membrane.

Sejnowski: Is there any advantage to them in being bigger cells?

Brenner: This is a fascinating area. Salamander cells are 30 times the volume of frog cells because they have 30 times more DNA. Thus you can make a salamander with only 3% of the number of cells of a frog. Function does not seem to change under these different conditions of scale and we need to make use of these natural experiments in our models.

Berridge: I would like to comment briefly on the significance of stochastic processes in signalling systems. Such stochastic events only become significant when they exceed the threshold for some cellular process. For example, cardiac cells produce random sparks, but they don't lead to contraction. However, if you add a low dose of caffeine or overload the cells with Ca^{2+}, then these stochastic events will spawn a regenerative Ca^{2+} wave, leading to a full contraction. We have to live with stochastic processes, but I think that most signalling systems have thresholds that filter out these random events.

Sejnowski: The promoter systems are often working down in that limit, where there are stochastic processes involved.

Prank: If there is an extracellular subthreshold stimulus, which won't be transduced across the cell membrane into intracellular Ca^{2+} oscillations, and if you add noise, the dynamic stimulus which is subthreshold becomes suprathreshold. This improves signal transduction, or makes signal transduction possible at all. This is an example of where stochasticity is not detrimental to signal transduction, but instead facilitates it.

Laughlin: Not only does having a threshold reduce the level of noise, but using positive feedback to drive the system to a saturated response also eliminates noise by making sure that all of the available signalling molecules are recruited. This use of positive feedback to increase reliability is an important design principle in cell signalling.

Sejnowski: That is also how digital computers work. They are making sure that they are only operating in regimes where it is saturated or not.

Laughlin: This is why there is a formal equivalence between the action potential, Ca^{2+} signalling and cAMP waves in *Dictyostelium*. They all point to the reliability of this method of signalling.

Iyengar: In simple systems this works, but what has always surprised me experimentally is that most of these signalling systems work at a very small fraction of their actual capability. You can get a lot more signal out of it if necessary, but a small fraction suffices.

Sejnowski: Part of this could be a sort of insurance policy.

Laughlin: If there are insufficient G protein molecules in the region of the cell where the receptor is activated, then it is likely that the signal will be lost. A minimum concentration of signalling molecules is required for a signal to be reliably transmitted and to achieve this over the entire cell requires a lot of molecules.

Sejnowski: That is another good point: spatial heterogeneity of signalling components. It may be that you need to have the concentration everywhere at a high value, not just where it is being produced.

Berridge: Dr Iyengar, what exactly do you mean when you talk about signalling systems operating at a small fraction of their capability?

Iyengar: I was thinking of more recent experiments that we have done with Ca^{2+}–calmodulin-activated protein kinase (CaMK) activation. Initially, when you surge in with Ca^{2+}, you can activate CaMK 20-fold and measure it in the tissue. But in the actual state at which we see long-term potentiation (LTP) consolidating, the activation is only sometimes 30–50% more than the threshold level.

Berridge: Is it 30–50% in those synaptic spines that have been activated, or are you looking at the whole cell?

Iyengar: It is the whole cell.

Berridge: The whole point about LTP is that it is input specific. You are only activating LTP in a very small proportion of the synapses of each neuron.

Iyengar: Not in the way that we do it. We use Kandel's so-called 'German' method, which delivers an intense stimulus. You can go back into those cells and give a second stimulus, which results in enormous CaMK activation. But only a small fraction of this is needed to get the physiological response.

Dolmetsch: One of the important things that people haven't mentioned yet is that there are often multiple thresholds for different kinds of events. For example, in the case of Ca^{2+}, very high Ca^{2+} kills a lot of cells. A low threshold of Ca^{2+} is required for other events. The challenge is to keep the signal between the two thresholds. One possibility is that if there is noise, all you need to do is elevate the average so that the noise barely peaks above the lower threshold. This ensures that you don't exceed the upper threshold, which is important if the two thresholds are subtly different. For example, if we have a lower threshold that has a memory, even if you only exceed the threshold periodically, you are still activating that signalling pathway, whereas you do not activate the upper pathway, because it requires more Ca^{2+} and has less memory because of its lower affinity.

Eichele: I find the issue of multiple thresholds very interesting. In developing systems, such as the specification of neurons in spinal cord, it has been shown that subtly different threshold concentrations of Sonic hedgehog (Shh) protein will evoke very different cell fates. This is not an artificial situation which is created *in vitro*, but is actually how ventral cell fates — say A and B — are defined in the spinal cord. I find it quite remarkable that the concentration differences in evoking fate A or B are small, just about five- to 10-fold. It is an interesting challenge to determine how this works. Threshold detection actually works through the same receptor system as far as we know.

Dolmetsch: There are two different Shh output systems. One possibility is that they differ not only in their affinity, but also in their on and off rates.

Eichele: As I said, there is only one receptor system for Shh.

Sejnowski: There is another factor: time. It is not just concentration of a signal, but also how long this concentration is maintained. Frequency is also important.

Eichele: In these experiments performed by Tom Jessell, there was no temporal variation. They just vary concentration. One of the key questions in development is how thresholds can specify the cell fates.

Brenner: Another key question concerns the number thresholds that can be read. Given a set of chemical systems, how many different levels can be generated and distinguished? All gradient theories need to deal with this problem. Different steady states cannot be sustained by one system, so a different mechanism is needed. It is interesting to analyse exactly how any particular system works. Thus in a chemical synapse, an impulse frequency is converted into a quantity of chemical transmitter, which is measured. Often, there are mechanisms to destroy or remove the transmitter, so that the currency of frequency is converted into the value of a pulse. We need to dissect all chemical communication systems in this way: do they count molecules, or do they measure different steady state levels? Is a signal transmitted as a pulse, or as a change of level? This is necessary before we can consider networks. One other point, in relation to the complexity of these systems is to recognize that they evolved by accretion. The fact that interleukin (IL)-5 and IL-4 are homologous proteins suggests that once there was just one protein and then the gene duplicated, to allow the evolution of different functions for them. Perhaps this was in response to new pathogens that had to be dealt with in a different way.

Laughlin: To evaluate signalling mechanisms we need to consider the magnitudes of the signalling processes. I was struck when Gregor Eichele said that a 10-fold change in the concentration of Shh was small. If one considers the numbers involved, this concentration change produces a large signal. Start with a modest number of molecules in a cell, say 100 bound to receptors at any one time. The binding of these molecules to receptors is subject to random fluctuations. Assuming a Poisson process, binding produces noise that fluctuates with a standard deviation of 10. When one increases the concentration 10-fold, to give 1000 bound molecules, the standard deviation of these fluctuations increases to 30. This 10-fold increase in concentration, from 100 to 1000, is 30 times larger than the standard deviation of the noise. This is a very robust and reliable signal. When you consider the relevant magnitudes, which in this case is numbers of molecules, you see why those of us who work with analogue signals in cells regard a 10-fold change as a large signal.

Eichele: In this case the Shh molecule can be associated with cholesterol moieties on the surface of the cell. This will elevate the concentration locally.

Laughlin: This signal, which you said was small, turns out to be an incredibly robust signal when you put the numbers in.

Eichele: The concentration of Shh is probably about 10^{-9} M.

Laughlin: None the less, if you are going to increase the concentration 10-fold you don't need many Shh molecules to get a very good signal.

Iyengar: Most of us who deal with membrane signalling find that numbers are not entirely useful to us. Counting the number of molecules is quite misleading when one of the reactants is in two dimensions and the other is in solution in three dimensions. Suzanne Scarlata has been measuring carefully what phospholipase C does, when it is dispersed or when it is in membranes. I have been struck by how different the numbers are when we have to make this factor correction.

Sejnowski: I wanted to turn the discussion towards the issue of evolution. Earlier on Sydney Brenner pointed out that there are some creatures with much more primitive immune systems, yet somehow they survive.

Brenner: These organisms without immune systems still have elaborate defence systems. *Drosophila* has a large number of genes that make lethal peptides. They don't have an adaptive immune system; they just react.

Sejnowski: What about plants?

Brenner: Plants respond to damage. They have chemical defences such as the phytoalexin response. Plants can respond by inducing enzymes such as chitinase as well as inhibitors of proteolytic enzymes. Defence is ancient.

Sejnowski: The reason I raised this is because there is a theory that all of these systems evolved as part of the warfare between hosts and pathogens. This raises the issue of sex: the rearrangement of genes in order to outrun the parasites. This has come up in the case of trying to model the brain. It is always a mistake to think that we are smarter than nature. According to Orgell's second law, 'nature is more clever than Leslie Orgell' (who in fact is a very clever person). In modelling the brain we often fall into the trap of assuming that we know what the function of a tissue is, that the purpose of vision is to recreate an internal model of the world, for example. Often these are implicit assumptions in the sense that no one questions them. What we discover after a lot of work is that nature isn't trying to do this for us, but the function is just a by-product or small part of what nature is trying to do. I wonder whether we are fooling ourselves in thinking that we can even guess what the goal of something is, given the fact that it evolved a long time ago for other purposes.

Brenner: I think there is another thing that we ought to watch out for: we have many combinations of signals, and it looks very complicated, but they may be present because of a 'don't care' condition. We often think that everything is specified and that genes are turned on and turned off in the combinations required. We should calculate the cost of evolving such elaborately specified regulation. If a gene happens to be turned on in a particular cell and has no effect, there will be no selection for repression of that instantiation. For example, encephalin is turned on in activated lymphocytes but does not do anything because a lymphocyte lacks the proteolytic secretory apparatus to process the product. Thus in this case, there would have been an evolutionary cost to develop a special control mechanism to turn it off in a lymphocyte, which it is not necessary to pay. I think many things will fall into this 'don't care' category. That is, you can have both IL-4 and IL-5; you may need IL-5 for eosinophils, but whether there is IL-4 or the two together doesn't matter because under those conditions there would be no advantage in having the unnecessary one turned off. This explanation of the apparent complexity can also account for many cases of apparent redundancy.

Dolmetsch: We need to devise some way of quantitating what the real cost is in producing a particular protein.

Brenner: Protein is not cheap. Regulation requires orthogonality of recognition, and that means we must cost new recognition elements. We know very little about this area. There are something like 1000 zinc finger regulatory genes in our genomes. This could generate a huge number of possible combinations; my belief is that most of these are not actually used.

Eichele: There is also a risk in turning genes off. This requires a change in the regulatory regions and such changes are not predictable in their effect on sites and levels of gene expression. In principle, this could even result in ectopic expression in cells that normally do not express a particular protein, a situation that could be detrimental to the organism.

Brenner: We are not 'running' evolution. All I am saying is that all natural selection must be treated simply on the grounds of reproductive success. Does IL-5 secretion do this? We test this by making a knockout of IL-5 in a mouse. This is the only experiment we can do to see whether fitness is decreased and, as you know, it is possible to knock out many genes in the mouse with no obvious decrease in fitness. Organisms cannot plan their genomes. A bacterium sitting the primitive ocean cannot say, 'I'd better not make this mutation in heat shock protein because I'm going to need actin in 2 billion years time'.

Kahn: If we evoke evolution of systems as a final guiding principle, we may oversimplify biology. Evolution can rediscover the same process in multiple ways. There are many species that fly that aren't derived linearly from one another. You can't say that because humans are more advanced than flies, that flying leads to disadvantage. We can't be strictly hierarchical in our evolutionary thinking. I am struck by the fact that in knockout experiments and in some cases tissue-specific knockouts, often when proteins are expressed in a tissue for which we thought they had no function they turn out to have a function. It is because we think in such limited roles of the function of a protein. Perhaps an interleukin in a non-lymphoid tissue may serve a completely different role than it would in the lymphoid tissue.

Sejnowski: It also might be the case that it doesn't have a function unless the cell is stressed or in an unusual condition.

Brenner: If we released every knockout mouse that we have made out into the wild, they most likely wouldn't survive, and come to think of it, normal mice would also not survive.

Fields: This discussion goes to the heart of the issue of the function of complexity: does it have a potential role in information processing, or is it a by-product of evolution. I guess we would all agree that it is not necessary, yet we find complex systems. What does complexity give us? Most of us would agree that it gives us resiliency through redundancy. If you lose a certain transcription factor or enzyme, you can still have a function. It also means that you can have more complex behaviour. You can induce LTP by many different types of stimuli, and therefore different mechanisms. This brings us to the question of what we should be modelling, and how we should go about it. Is the best approach to use reaction kinetics and build our way up through this system, or do we need a more unifying idea, such as information or energy flow through trophic systems?

Brenner: I know one approach that will fail, which is to start with genes, make proteins from them and to try to build things bottom–up.

The versatility and complexity of calcium signalling

Michael J. Berridge

Laboratory of Molecular Signalling, The Babraham Institute, Babraham Hall, Cambridge CB2 4AT, UK

Abstract. Ca^{2+} is a universal second messenger used to regulate a wide range of cellular processes such as fertilization, proliferation, contraction, secretion, learning and memory. Cells derive signal Ca^{2+} from both internal and external sources. The Ca^{2+} flowing through these channels constitute the elementary events of Ca^{2+} signalling. Ca^{2+} can act within milliseconds in highly localized regions or it can act much more slowly as a global wave that spreads the signal throughout the cell. Various pumps and exchangers are responsible for returning the elevated levels of Ca^{2+} back to the resting state. The mitochondrion also plays a critical role in that it helps the recovery process by taking Ca^{2+} up from the cytoplasm. Alterations in the ebb and flow of Ca^{2+} through the mitochondria can lead to cell death. A good example of the complexity of Ca^{2+} signalling is its role in regulating cell proliferation, such as the activation of lymphocytes. The Ca^{2+} signal needs to be present for over two hours and this prolonged period of signalling depends upon the entry of external Ca^{2+} through a process of capacitative Ca^{2+} entry. The Ca^{2+} signal stimulates gene transcription and thus initiates the cell cycle processes that culminate in cell division.

2001 Complexity in biological information processing. Wiley, Chichester (Novartis Foundation Symposium 239) p 52–67

The universality of Ca^{2+} as an intracellular messenger depends upon its enormous versatility. Many of the molecular components of the Ca^{2+} signalling system have multiple isoforms that can be mixed and matched to create a wide range of spatial and temporal signals. Ca^{2+} can operate within milliseconds in highly localized regions or it can act much more slowly as global waves of Ca^{2+} spreading throughout the cell or through large groups of cells. This versatility, which is exploited to control processes as diverse as fertilization, cell proliferation, development, secretion, chemotaxis, learning and memory must all be accomplished within the context of Ca^{2+} being a highly toxic ion. If its normal spatial and temporal boundaries are exceeded, this deregulation of Ca^{2+} signalling results in cell death through both necrosis and apoptosis. The aim of this review is twofold: I will first describe the complex nature of Ca^{2+} signalling

52

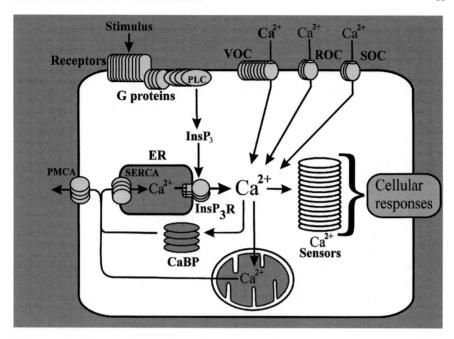

FIG. 1. The Ca^{2+} signalling toolkit. All of the molecular components regulating the Ca^{2+} signalling pathway are composed of multiple components, often closely related isoforms with subtly different properties. The duplication shown on the figure illustrates the degree of this diversity. Inositol 1,4,5-trisphosphate (InsP$_3$), which is generated by phospholipase C (PLC), acts on InsP$_3$ receptors (InsP$_3$R) located on the endoplasmic reticulum (ER). Ca^{2+} enters the cell through multiple isoforms of voltage-operated channels (VOCs), receptor-operated channels (ROCs) and store-operated channels (SOCs). Various Ca^{2+}-binding proteins (CaBPs) buffer Ca^{2+} both in the cytoplasm and within the lumen of the ER. Ca^{2+} is pumped out of the cell by exchangers and plasma membrane Ca^{2+}-ATPases (PMCA) or into the ER by sarcoendoplasmic reticulum Ca^{2+}-ATPases (SERCA).

and then consider how this messenger system functions in cell proliferation and cell death.

The Ca^{2+} signalling network

The hallmark of Ca^{2+} signalling is its complexity. One manifestation of this is the existence of two separate sources of Ca^{2+}, which can be derived from either internal stores or by uptake from the external medium (Fig. 1). Different channels and pumps regulate each source. Signalling begins when the external stimulus binds to receptors that either activate channels in the plasma membrane or generate Ca^{2+}-mobilizing signals that release Ca^{2+} from the internal stores. The Ca^{2+} that flows into the cytoplasm functions as a messenger to stimulate numerous

Ca^{2+}-sensitive processes. Finally, there are OFF mechanisms, composed of pumps and exchangers, which remove Ca^{2+} from the cytoplasm to restore the resting state. Most of the processes of the signalling pathway are carried out by different components, which means that each cell has access to a diverse molecular toolkit (Fig. 1). By mixing and matching all the available possibilities, cells can create Ca^{2+} signals with widely different spatial and temporal properties.

Generation of Ca^{2+} signals

There are families of Ca^{2+} entry channels defined by the way in which they are activated. We know most about voltage-operated channels (VOCs) of which there are at least 10 types (Fig. 1) with subtly different properties. Ca^{2+} can also enter cells through receptor-operated channels (ROCs) and through store-operated channels (SOCs). There is considerable debate as to how empty stores can activate channels in the plasma membrane. Recent evidence has begun to support a conformational-coupling mechanism, which proposed that the inositol-1,4,5-trisphosphate ($InsP_3$) receptors in the plasma membrane are directly coupled to the SOCs in the plasma membrane (Berridge et al 2000). There is considerable interest in these SOCs since they provide the Ca^{2+} signal that controls cell proliferation (see later).

Signal Ca^{2+} is also derived from the internal stores using channels regulated by Ca^{2+}-mobilizing messengers, such as $InsP_3$ that diffuses into the cell to engage the $InsP_3$ receptors ($InsP_3Rs$) that release Ca^{2+} from the endoplasmic reticulum (ER). Cyclic ADP ribose (cADPR) acts by releasing Ca^{2+} from ryanodine receptors (RYRs). Sphingosine-1-phosphate (S1P) and nicotinic acid dinucleotide phosphate (NAADP) release Ca^{2+} by binding to channels that have yet to be characterized.

Most attention has focused on the $InsP_3Rs$ and the RYRs, which are regulated by a number of factors — the most important of which is Ca^{2+} itself. For example, the $InsP_3Rs$ have a bell-shaped Ca^{2+} dependence in that low concentrations (100–300 nM) are stimulatory but above 300 nM, Ca^{2+} becomes inhibitory and acts to switch the channel off. Once the receptor binds $InsP_3$, it becomes sensitive to the stimulatory action of Ca^{2+}. In the same way, cADPR increases the Ca^{2+} sensitivity of the RYRs. The $InsP_3Rs$ and the RYRs have a mechanism of *Ca^{2+}-induced Ca^{2+} release* (CICR) and this autocatalytic process enables individual channels to communicate with each other to establish highly coordinated Ca^{2+} signals often organized into propagating waves. The main function of the Ca^{2+}-mobilizing messengers, therefore, is to alter the sensitivity of the $InsP_3Rs$ and RYRs to this stimulatory action of Ca^{2+}.

These different Ca^{2+}-mobilizing messengers often coexist in cells where they seem to be controlled by different receptors. For example, in the exocrine pancreas, acetylcholine receptors act through InsP$_3$ whereas CCK receptors employ NAADP and cADPR (Cancela et al 1999). Similarly, human SH-SY5Y cells have acetylcholine receptors linked through InsP$_3$ while lysophosphatidic acid (LPA) acts through S1P (Young et al 1999). The complexity of the signalling network is thus enhanced by having different Ca^{2+}-mobilizing messengers linked to separate input signals.

Ca^{2+}-sensitive processes

Various Ca^{2+}-sensitive processes translate Ca^{2+} signals into cellular responses. There are a large number of Ca^{2+}- binding proteins, which can be divided into Ca^{2+} sensors and Ca^{2+} buffers. The Ca^{2+} sensors respond to the increase in Ca^{2+} by activating a wide range of responses. Classical examples of sensors are troponin C (TnC) and calmodulin, which have four E-F hands that bind Ca^{2+} and undergo a pronounced conformational change to activate a variety of downstream effectors. TnC has a somewhat limited function of controlling the interaction of actin and myosin during the contraction cycle of cardiac and skeletal muscle. By contrast, calmodulin is used much more generally to regulate many different processes such as the contraction of smooth muscle, cross-talk between signalling pathways, gene transcription, ion channel modulation and metabolism. The same cell can use different detectors to regulate separate processes. In skeletal muscle, for example, TnC regulates contraction whereas calmodulin stimulates phosphorylase thereby ensuring an increase in ATP production.

In addition to the above proteins, which have a more general function, there are a large number of Ca^{2+}-binding proteins designed for more specific functions. For example, synaptotagmin is associated with membrane vesicles and is responsible for mediating exocytosis. A large family of S100 Ca^{2+}-binding proteins seems to be particularly important in cell proliferation and have been implicated in cancer growth and metastasis. For example, human chromosome Iq21 has a cluster of approximately 10 S100 genes that are differentially expressed in neoplastic tissues. Melanoma cells overexpress S100B and antibodies against this Ca^{2+}-binding protein are used for tumour typing and diagnosis of melanoma. S100B can activate a nuclear serine/threonine protein kinase (Millward et al 1998) and can cooperate with protein kinase C to induce the translocation of p53 early in the G1 phase of the cell cycle (Scotto et al 1999).

Once Ca^{2+} has carried out its signalling functions, it is rapidly removed from the cytoplasm by various pumps and exchangers located both on the plasma membrane and on the internal stores (Fig. 1). The mitochondrion is another important component of the OFF mechanism in that it sequesters Ca^{2+} rapidly during the recovery phase and then slowly releases it back when the cell is at rest. In order to synthesize ATP, the mitochondrion extrudes protons to create the electrochemical gradient that is used to synthesize ATP. Exactly the same gradient is used to drive Ca^{2+} uptake through a uniporter which functions much like a channel. The mitochondrion has a large capacity to accumulate Ca^{2+}. Once the cytosolic level of Ca^{2+} has returned to its resting level, a Na^+/Ca^{2+} exchanger transfers the large load of Ca^{2+} back into the cytoplasm where it is once again returned to the ER or removed from the cell. In addition to this slow efflux pathway, Ca^{2+} can also leave through a permeability transition pore (PTP). This PTP may have two functional states. First, there is a low conductance state that acts reversibly, allowing mitochondria to become excitable and thus contributing to the generation of Ca^{2+} waves (Ichas et al 1997). Second, there is an irreversible high conductance state of the PTP that has a dramatic effect on the mitochondrion in that it collapses the transmembrane potential and leads to the release of cyctochrome c and the initiation of apoptosis.

During normal signalling, therefore, there is a continuous ebb and flow of Ca^{2+} between the ER and the mitochondria. At the onset of each spike, a small bolus of Ca^{2+} is released to the cytoplasm and some of this signal enters the mitochondria where it has a temporary residence before being returned to the ER. Mitochondria contribute to the onset of apoptosis if this normal exchange of Ca^{2+} with the ER is distorted.

The apoptosis regulatory proteins that function either as death antagonists (Bcl2 and $BclX_L$) or death agonists (Bax, Bak and Bad), may exert some of their actions by interfering with the Ca^{2+} dynamics of these two organelles. For example, Bax and Bad accelerate the opening of the voltage-dependent anion channel, which is part of the PTP, and thus contribute to the release of cytochrome c (Shimizu et al 1999). On the other hand, Bcl2 and $BclX_L$ seem to act by blocking Ca^{2+}-induced apoptosis, enabling the mitochondria to cope with large loads of Ca^{2+} (Zhu et al 1999). Bcl2 is also present on the ER where it acts to enhance the store of Ca^{2+} (Zhu et al 1999) perhaps by up-regulating the expression of Ca^{2+} pumps (Kuo et al 1998).

Spatial and temporal aspects of Ca^{2+} signalling

Much of the versatility of Ca^{2+} signalling arises from the way that it is presented in both time and space. Our understanding of the spatial aspects has increased enormously due to advances in imaging technology that have enabled us to visualize the elementary events of Ca^{2+} signalling. These elementary events are the basic building blocks of Ca^{2+} signals in that they represent the Ca^{2+} that results from the opening of either single or small groups of channels. They have been described most extensively for the channels that release Ca^{2+} from the internal stores. Whether or not these channels open to release Ca^{2+} is determined by their degree of excitability, which is controlled by a number of factors. As described earlier, the primary determinant for the InsP$_3$Rs is InsP$_3$ whereas the RYRs are sensitive to cADPR. Both channels are also sensitive to the degree of Ca^{2+} loading in the store.

At low levels of stimulation, the level of excitability is such that individual RYRs or InsP$_3$Rs open; such events have been recorded as quarks or blips, respectively. They may be considered as the fundamental events that form the basis of most Ca^{2+} signals. These single-channel events are rare and the more usual event is somewhat larger, resulting from the coordinated opening of small groups of InsP$_3$Rs or RYRs known as puffs or sparks, respectively. Sparks were first described in cardiac cells where they arise from a group of RYR2 channels opening in response to Ca^{2+} entering through L channels. Puffs have a wide range of amplitudes suggesting that there are variable numbers of InsP$_3$Rs within each cluster. These sparks and puffs are the elementary events of Ca^{2+} signalling that contribute to the intracellular waves that sweep through cells to create global Ca^{2+} signals. When gap junctions connect cells, such intracellular waves can spread to neighbouring cells thus creating intercellular waves capable of coordinating the activity of large groups of cells.

In addition to creating global responses, these elementary events have another important function in that they can carry out signalling processes within highly localized cellular domains. A classic example is the process of exocytosis at synaptic endings where N- or P/Q-type VOCs create a local pulse of Ca^{2+} to activate synaptotagmin to trigger vesicle release. Sparks located near the plasma membrane of excitable cells activate Ca^{2+}-sensitive K$^+$ channels bringing about membrane hyperpolarization, which can regulate the excitability of neurons or the contractility of smooth muscle cells. In HeLa cells, Ca^{2+} puffs are concentrated around the nucleus where they feed Ca^{2+} directly into the nucleoplasm (Lipp et al 1997). Finally, as mentioned earlier, the mitochondria located near the sites of elementary events take up Ca^{2+} rapidly and this stimulates mitochondrial metabolism to increase the formation of ATP.

In addition to these spatial variations, there are also marked differences in the temporal aspect of Ca^{2+} signalling. More often than not, Ca^{2+} signals are presented as brief spikes. In some cases, individual spikes are sufficient to trigger a cellular response as occurs during contraction of skeletal or cardiac muscle or the release of synaptic vesicles by exocytosis. When longer periods of signalling are necessary, such spikes are repeated to give oscillations with widely differing frequencies. Periods within the 1–60 second range are found in the pancreas and liver, but much longer periods of 1–5 minutes have been recorded in mammalian eggs following fertilization. A Ca^{2+} oscillator that initiates mitosis during the cell cycle has an even longer period of signalling of approximately 24 hours.

The mitotic Ca^{2+} oscillator is particularly interesting because it is an integral component of the control mechanisms that regulate the cell cycle. The latter is an orderly programme of events controlled by two linked oscillators: a cell cycle oscillator and the Ca^{2+} oscillator (Swanson et al 1997). The former depends upon the synthesis and periodic proteolysis of various cyclins at specific points during the cell cycle. The Ca^{2+} oscillator, based on the periodic release of stored Ca^{2+}, is responsible for initiating specific events associated with mitosis such as nuclear envelope breakdown (NEBD), anaphase and cell cleavage. As the one-cell mouse embryo approaches its first mitosis, there are a series of spontaneous Ca^{2+} transients responsible for triggering various events during mitosis such as NEBD, anaphase and cleavage to the two-cell stage (Chang & Meng 1995). Just what drives the Ca^{2+} oscillator is a mystery but there are indicators that it depends upon the periodic elevation of $InsP_3$. In the case of the sea urchin, the level of $InsP_3$ is increased at distinct points during mitosis such as NEBD, anaphase and cleavage, at the time of each spontaneous Ca^{2+} transient (Ciapa et al 1994).

When cells need to be activated for prolonged periods, a single Ca^{2+} spike is not sufficient and is replaced by Ca^{2+} oscillations. Cells respond to changes in stimulus intensity by varying spike frequency. Such frequency-modulated signalling is used to control processes such as liver metabolism, smooth muscle contractility and differential gene transcription, especially in developing systems. For example, presenting Ca^{2+} in the form of spikes was more effective in initiating gene expression than a steady maintained level of Ca^{2+} (Li et al 1998). A low frequency of spiking activated NF-κB, whereas higher frequencies were necessary to switch on NF-AT and Oct (Dolmetsch et al 1998). Such oscillatory activity is particularly important for the development of both neural and muscle cells (Buonanno & Fields 1999). In *Xenopus*, spontaneous Ca^{2+} spikes produced by RYRs during a narrow developmental window are responsible for the differentiation of myocytes into somites (Ferrari et al 1998). Neural development is also mediated by Ca^{2+} spikes that control process such as differentiation (Gu & Spitzer 1997, Carey & Matsumoto 1999), the behaviour of growth cones (Gomez & Spitzer 1999) and the establishment of the specific connections within neural circuits (Feller 1999).

In order to use such a frequency-modulated signalling system, cells have evolved sophisticated 'molecular machines' for decoding such frequency encoded Ca^{2+} signals. The two Ca^{2+}-sensitive proteins that seem to play a role in decoding are CaM kinase II (DeKoninck & Schulman 1998) and protein kinase C (Oancea & Meyer 1999).

Cell proliferation

A good example of the complexity of Ca^{2+} signalling is its role in regulating cell proliferation. Once cells have differentiated to perform specific functions, they usually stop proliferating. In many cases, however, such differentiated cells maintain the option of returning to the cell cycle and this usually occurs in response to growth factors. For example, lymphocytes proliferate rapidly in response to antigens, smooth muscle cells respond to growth factors such as PDGF at the sites of wounds and astrocytes are stimulated to grow at sites of brain injury. In many of these examples, Ca^{2+} is one of the key regulators of cell proliferation, where it functions in conjunction with other signalling pathways such as those regulated through MAP kinase and phosphatidylinositol-3 kinase (PI 3-K) (Lu & Means 1993, Berridge 1995). The function of Ca^{2+} in regulating cell proliferation is well illustrated in lymphocytes responding to antigen. Figure 2 attempts to summarize all the signalling elements that are used by a T cell as it responds to the arrival of an antigen. In this case, the antigen functions as a 'growth factor' that binds to the T cell receptor to initiate the assembly of a 'supramolecular activation cluster' (Monks et al 1998) containing scaffolding and signal transducing elements. The latter function to relay information into the nucleus using various signalling cassettes. The cassettes linked to phospholipase C (PLC)γ1, which produces both diacylglycerol (DAG) and InsP$_3$, are particularly important and frequently associated with the action of growth factors and have been implicated in cell transformation. In fact, PLCγ has been referred to as a malignancy linked signal transducing enzyme (Yang et al 1998) and its overexpression will promote transformation and tumorigenesis in NIH 3T3 cells (Smith et al 1998). The InsP$_3$ formed by PLCγ1 releases Ca^{2+} from the internal store, which then promotes entry of external Ca^{2+} through a SOC.

When used for controlling cell proliferation, this Ca^{2+} signalling pathway needs to be active for a prolonged period — two hours in the case of lymphocytes. Since the stores have a very limited capacity, this prolonged period of Ca^{2+} signalling is critically dependent on this influx of external Ca^{2+}. There are two modulatory mechanisms that function to maintain Ca^{2+} signalling (Fig. 3). The first is an example of the cross talk between signalling pathways and concerns the ability of PI 3-K to stimulate PLCγ1 to maintain the supply of InsP$_3$ (Scharenburg & Kinet 1998). Formation of the lipid second messenger PIP$_3$ activates the non-receptor tyrosine kinase Btk that then phosphorylates and activates PLCγ1. The tumour suppressor PTEN, which acts as a 3-phosphatase to lower the level of phosphatidylinositol-3,4,5-trisphosphate (PIP$_3$), reduces both the level of InsP$_3$ and the influx of external Ca^{2+} (Morimoto et al 2000). The second is the activation of potassium channels that serve to hyperpolarize the membrane which is essential to maintain the entry of external Ca^{2+} (Lewis & Cahalan 1995). For example, cell proliferation is regulated by IK$_{Ca}$, which is inhibited by charybdotoxin and iberiotoxin. The net effect of these two mechanisms is to ensure a continuous influx of external Ca^{2+}, which seems to be one of the principle early signals to promote cell proliferation. A Ca^{2+} influx inhibitor carboxy-amidotriazole can prevent cell proliferation and has been used in clinical trials to control refractory cancers (Kohn et al 1996).

T-CELL ACTIVATION

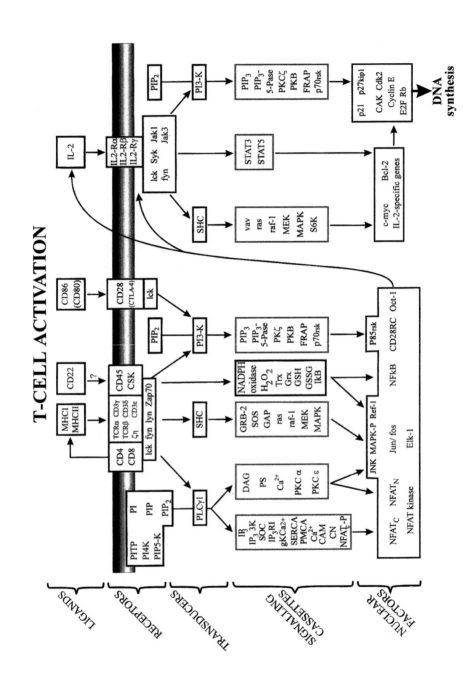

DNA
synthesis

LIGANDS

RECEPTORS

TRANSDUCERS

SIGNALLING
CASSETTES

NUCLEAR
FACTORS

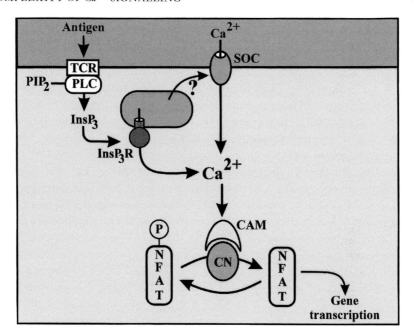

FIG. 3. The role of Ca^{2+} in lymphocyte activation. The antigen activates the T cell receptor (TCR), which stimulates phospholipase C (PLC) to hydrolyse phosphatidylinositol-4,5-bisphosphate (PIP$_2$), to release inositol-1,4,5-trisphosphate (InsP$_3$) to the cytosol. InsP$_3$ releases Ca^{2+} from the internal store, which then sends an unknown signal to the store-operated channels (SOCs) in the plasma membrane. Ca^{2+} acts through calmodulin (CAM) to stimulate calcineurin (CN) which dephosphorylates the nuclear factor of activated T cells (NFAT) enabling it to enter the nucleus to initiate gene transcription.

The main function of Ca^{2+} is to activate transcription factors either in the cytoplasm (NF-AT, NF-κB) or within the nucleus (CREB). The role of Ca^{2+} in stimulating gene transcription is very similar in neurons undergoing learning as it is in cells being induced to grow. An increase in Ca^{2+} is one of the signals capable of bringing about the hydrolysis of the inhibitory IκB subunit allowing the active NF-κB subunit to enter the nucleus. Perhaps the most important action of Ca^{2+} is to stimulate calcineurin to dephosphorylate NF-AT, which then enters the nucleus (Fig. 3) (Crabtree 1999). As soon as Ca^{2+} signalling ceases,

---

FIG. 2. A spatiotemporal map of T cell activation. The spatial aspect (i.e. from top to bottom) concerns the way in which the antigen binding to the complex T cell receptor activates a number of signalling cassettes that transfer information from the plasma membrane to the nucleus. The temporal aspect (i.e. from left to right) deals with the flow of information through the sequential signalling elements that occurs during the protracted G1 period of the cell cycle and culminates in the activation of either proliferation or apoptosis.

kinases in the nucleus rapidly phosphorylate NF-AT which then leaves the nucleus and transcription ceases. The prolonged period of Ca^{2+} signalling that is required for proliferation to occur is thus necessary to maintain the transcriptional activity of NF-AT. Transcription is inhibited in mutants with defective SOCs that cannot sustain Ca^{2+} signalling (Timmerman et al 1996). Likewise, the immuno-suppressant drugs cyclosporin A and FK506 prevent transcription by inhibiting the action of calcineurin. In contrast to the previous two transcription factors that are activated within the cytoplasm, CREB is a nuclear Ca^{2+}-responsive element, which is phosphorylated by CaMKII and CaMKIV. In addition, Ca^{2+} acting within the nucleus is also responsible for stimulating the Ca^{2+}-sensitive transcriptional coactivator CREB-binding protein (CBP) (Hardingham et al 1999). A CaM inhibitory peptide targeted to the nucleus was able to block DNA synthesis and cell cycle progression thus emphasizing the importance of a nuclear Ca^{2+} signal for cell proliferation (Wang et al 1996). These transcription factors activate a large number of target genes, some code for progression factors such as the interleukin 2 system responsible for switching on DNA synthesis whereas others produce components such as Fas and the Fas ligand that are responsible for apoptosis (Fig. 2). Ca^{2+} thus plays a central role in putting in place the signalling systems that enable cells to decide whether to grow or to die.

References

Berridge MJ 1995 Calcium signalling and cell-proliferation. Bioessays 17:491–500
Berridge MJ, Lipp P, Bootman MD 2000 Signal transduction. The calcium entry pas de deux. Science 287:1604–1605
Buonanno A, Fields RD 1999 Gene regulation by patterned electrical activity during neural and skeletal muscle development. Curr Opin Neurobiol 9:110–120
Cancela JM, Churchill GC, Galione A 1999 Coordination of agonist-induced Ca^{2+}-signalling patterns by NAADP in pancreatic acinar cells. Nature 398:74–76
Carey B, Matsumoto SG 1999 Spontaneous calcium transients are required for neuronal differentiation of murine neural crest. Dev Biol 215:298–313
Chang DC, Meng CL 1995 A localized elevation of cytosolic-free calcium is associated with cytokinesis in the zebrafish embryo. J Cell Biol 131:1539–1545
Ciapa B, Pesando D, Wilding M, Whitaker M 1994 Cell-cycle calcium transients driven by cyclic changes in inositol trisphosphate levels. Nature 368:875–878
Crabtree GR 1999 Generic signals and specific outcomes: signaling through Ca^{2+}, calcineurin, and NF-AT. Cell 96:611–614
De Koninck P, Schulman H 1998 Sensitivity of CaM kinase II to the frequency of Ca^{2+} oscillations. Science 279:227–230
Dolmetsch RE, Xu KL, Lewis RS 1998 Calcium oscillations increase the efficiency and specificity of gene expression. Nature 392:933–936
Feller MB 1999 Spontaneous correlated activity in developing neural circuits. Neuron 22:653–656
Ferrari MB, Ribbeck K, Hagler DJ, Spitzer NC 1998 A calcium signaling cascade essential for myosin thick filament assembly in Xenopus myocytes. J Cell Biol 141:1349–1356

Gomez TM, Spitzer NC 1999 *In vivo* regulation of axon extension and pathfinding by growth-cone calcium transients. Nature 397:350–355

Gu XN, Spitzer NC 1997 Breaking the code: regulation of neuronal differentiation by spontaneous calcium transients. Dev Neurosci 19:33–41

Hardingham GE, Chawla S, Cruzalegui FH, Bading H 1999 Control of recruitment and transcription-activating function of CBP determines gene regulation by NMDA receptors and L-type calcium channels. Neuron 22:789–798

Ichas F, Jouaville LS, Mazat JP 1997 Mitochondria are excitable organelles capable of generating and conveying electrical and calcium signals. Cell 89:1145–1153

Kohn EC, Reed E, Sarosy G et al 1996 Clinical investigation of a cytostatic calcium influx inhibitor in patients with refractory cancers. Cancer Res 56:569–573

Kuo TH, Kim HRC, Zhu LP, Yu YJ, Lin HM, Tsang W 1998 Modulation of endoplasmic reticulum calcium pump by Bcl-2. Oncogene 17:1903–1910

Lewis RS, Cahalan MD 1995 Potassium and calcium channels in lymphocytes. Annu Rev Immunol 13:623–653

Li WH, Llopis J, Whitney M, Zlokarnik G, Tsien RY 1998 Cell-permeant caged InsP$_3$ ester shows that Ca^{2+} spike frequency can optimize gene expression. Nature 392:936–941

Lipp P, Thomas D, Berridge MJ, Bootman MD 1997 Nuclear calcium signalling by individual cytoplasmic calcium puffs. EMBO J 16:7166–7173

Lu KP, Means AR 1993 Regulation of the cell cycle by calcium and calmodulin. Endocr Rev 14:40–58

Millward TA, Heizmann CW, Schafer BW, Hemmings BA 1998 Calcium regulation of Ndr protein kinase mediated by S100 calcium-binding proteins. EMBO J 17:5913–5922

Monks CRF, Freiberg BA, Kupfer H, Sciaky N, Kupfer A 1998 Three-dimensional segregation of supramolecular activation clusters in T cells. Nature 395:82–86

Morimoto AM, Tomlinson MG, Nakatani K, Bolen JB, Roth RA, Herbst R 2000 The MMAC1 tumor suppressor phosphatase inhibits phospholipase C and integrin-linked kinase-activity. Oncogene 19:200–209

Oancea E, Meyer T 1998 Protein kinase C as a molecular machine for decoding calcium and diacylglycerol signals. Cell 95:307–318

Scharenberg AM, Kinet JP 1998 PtdIns-3,4,5-P$_3$: a regulatory nexus between tyrosine kinases and sustained calcium signals. Cell 94:5–8

Scotto C, Delphin C, Deloulme JC, Baudier J 1999 Concerted regulation of wild-type p53 nuclear accumulation and activation by S100B and calcium-dependent protein kinase C. Mol Cell Biol 19:7168–7180

Shimizu S, Narita M, Tsujimoto Y 1999 Bcl-2 family proteins regulate the release of apoptogenic cytochrome c by the mitochondrial channel VDAC. Nature 399:483–487

Smith MR, Court DW, Kim HK et al 1998 Overexpression of phosphoinositide-specific phospholipase Cγ in NIH 3T3 cells promotes transformation and tumorigenicity. Carcinogenesis 19:177–185

Swanson CA, Arkin AP, Ross J 1997 An endogenous calcium oscillator may control early embryonic division. Proc Natl Acad Sci USA 94:1194–1199

Timmerman LA, Clipstone NA, Ho SN, Northrop JP, Crabtree GR 1996 Rapid shuttling of NF-AT in discrimination of Ca^{2+} signals and immunosuppression. Nature 383:837–840

Wang JH, Moreira KM, Campos B, Kaetzel MA, Dedman JR 1996 Targeted neutralization of calmodulin in the nucleus blocks DNA synthesis and cell cycle progression. Biochim Biophys Acta 1313:223–228

Yang H, Shen F, Herenyiova M, Weber G 1998 Phospholipase C (EC 3.1.4.11): a malignancy linked signal transduction enzyme. Anticancer Res 18:1399–1404

Young KW, Challiss RAJ, Nahorski SR, Mackrill JJ 1999 Lysophosphatidic acid-mediated Ca^{2+} mobilization in human SH-SY5Y neuroblastoma cells is independent of phosphoinositide signalling, but dependent on sphingosine kinase activation. Biochem J 343:45–52
Zhu LP, Ling S, Yu XD et al 1999 Modulation of mitochondrial Ca^{2+} homeostasis by Bcl-2. J Biol Chem 274:33267–33273

DISCUSSION

Fields: You said something that appears contradictory. You said that subthresold events such as low agonist concentrations lead to local Ca^{2+} changes, and then higher intensity stimulation leads to a global Ca^{2+} response. Then you also told us that the local Ca^{2+} changes in spines are important for long-term potentiation (LTP). I would be interested in your comments on this. The normal stimulus that induces LTP causes the neuron to fire action potentials and therefore causes a global change in Ca^{2+}. How can the importance of local Ca^{2+} change be reconciled with situations where the stimulus would produce a global Ca^{2+} change?

Berridge: In the case of LTP one has to be very careful in terms of understanding local versus global effects. In the case of the spine, Ca^{2+} is input specific in the sense that it is elevated in only those synapses that are active. In addition, you can get global changes, but my feeling is that the concentration caused by a global Ca^{2+} change within the whole dendritic tree will probably not reach threshold to modify individual spines. Neurons have an enormous concentration of Ca^{2+} buffers and they vary in the proportion of these buffers that are expressed. One of the functions of such buffers is to dampen out Ca^{2+} signals. Every time the neuron fires, as part of its information-processing role, there is a back-propagating action potential that spreads into the dendritic tree, resulting in a global Ca^{2+} change. You don't want to modify your synapses every time you are processing information. Although this is a global Ca^{2+} change, I would argue that the buffers ensure that the concentration is relatively low within individual spines. However, if you have a back-propogated action potential occurring in conjunction with the activation of a synapse, then there will be a much larger but localized elevation of Ca^{2+} within the spine.

Fields: Are there measurements that indicate that the Ca^{2+} rise is insulated or augmented in the spine?

Berridge: The measurements are only really just starting, but we already have recordings to show that Ca^{2+} signals can be restricted to individual spines. It is amazing that we are actually able to measure the Ca^{2+} in these spines: the estimate is that at rest there are only six free Ca^{2+} ions in each spine.

Sejnowski: We have done simulation of Ca^{2+} entry into the spines of pyramidal cells. You have to be careful interpreting these pictures, because the Ca^{2+} indicator is itself a buffer.

Noble: I love the emphasis on beauty rather complexity. It is the unravelling of these beautiful systems that is the great joy to those of us who are trying to work at higher levels. Adding to your versatility, there is at least one cell for which the *amplitude* of the global signal is variable in an important way, which is the heart. One of the reasons I say this is connected with my comment on the joy of unravelling complex systems: it was actually quite difficult to get a graded release in modelling the Ca^{2+} release mechanism in the heart. I would even go so far to say that although we have now got models of this, we still don't fully understand why it is as graded as it is.

Berridge: Recent studies on Ca^{2+} sparks in cardiac cells have provided an explanation for such graded responses. This has been one of the unsolved problems in physiology: how can a process of Ca^{2+}-induced Ca^{2+} release (CICR), which is a positive feedback mechanism, generate a graded release of Ca^{2+}? One of the ideas is that the gradation may actually depend upon the variable recruitment of these elementary events.

Noble: Exactly. You may need to bring in what you call the 'physiological toolkit'. Trying to model with just one variable of free Ca^{2+} is only possible provided you make Ca^{2+} do two things: activate the release and also inactivate it with another time course. People who have tried to isolate that inactivation process by breaking the system down into its component bits can't find it. There has to be something else, and I have a strong suspicion it may be in what you call the physiological toolkit: the way Ca^{2+} is located in complex physiological spaces and structures.

Berridge: To expand on this idea of the physiological toolkit, it looks very much as if at low depolarizations, relatively few of these individual events are recruited. Each of these spark sites is an autonomous unit that fires independently of the others. Each unit is using CICR, but because they are separated from each other Ca^{2+} doesn't spread to neighbouring sites. By varying the level of depolarization, you can recruit variable numbers of these individual events. By having autonomous units it is possible to get a graded response. It is a very elegant solution.

Schultz: You mentioned the fact that cADP ribose stimulates the Ca^{2+} release and that cGMP stimulates cADP ribose formation. In which systems is this control important? I know systems in which cGMP blocks Ca^{2+} release and others where it stimulates Ca^{2+} influx, but I don't really know of a system in which cGMP would stimulate Ca^{2+} release.

Berridge: That's a good point. This was a very old slide which came from work on the sea urchin, where there seems to be some indication of cGMP playing a role in fertilization. The idea is that fertilization may generate cADP ribose, which in the sea urchin is responsible for activating Ca^{2+} release from the ER. In other cell types, there is some uncertainty concerning cADP ribose with regard to its precise function. There's evidence that it is playing an important role in the pancreas, which is activated by different hormonal systems. Acetylcholine acts through the $InsP_3$ system, whereas CCK seems to use cADP ribose.

Sejnowski: I liked your idea about looking at spatial scales. You used the micron scale as a convenient one. I'd like to suggest that there is actually a sub-micron scale that is equally important, especially in synapses, because of the fact that much of the machinery of the receptors and the Ca^{2+}–calmodulin complexes are right there under the plasma membrane, organized in a very precise way. In fact, these are little machines: they are really complexes. They are positioned such that when Ca^{2+} does enter, say through the NMDA receptor, it is at a very high concentration, briefly and locally. It could be that there is an even more precise molecular machinery.

Berridge I agree that we need to study more closely the precise morphology and molecular organization of the spine. There's some intriguing evidence coming out in terms of the variation between, for example, the CA1 neuron and the Purkinje cell in terms of the distribution of the metabotropic receptors, the G_q/G_{11} transducers and various phospholipase C isoforms.

Segel: You mentioned that contraction and relaxation occur together in smooth muscle. I have thought for a long time that there must be some sort of system that makes smooth muscle contract smoothly, in the sense of coordinating all the sub-parts of it. Has that been studied? Is there such a system where you would need both contraction and relaxation together to make everything work in a coordinated fashion?

Berridge: There is not just one type of smooth muscle, but instead there is an enormous variety of types. A lot of these smooth muscle cells function in a tonic state of contraction: they are poised between contraction and relaxation. This dynamic equilibrium between relaxation and contraction may be governed by the spatial organization of the Ca^{2+} signalling system. Günter Schultz is an expert on smooth muscle: perhaps he would like to comment on this?

Schultz: With regard to the tonic aspect of smooth muscle contraction, the Ca^{2+}-independent pathway causes contraction via G_{12} and G_{13} signalling to Rho and Rho kinase, and inhibition of myosin phosphatase. This is an important aspect, contributing as much as the Ca^{2+} part does to the overall contraction.

Laughlin: If one sat down with a piece of paper and decided to design a signalling system in a cell which depended on the propagation of some signalling molecule, one probably would not choose a molecule that was very heavily buffered. Because Ca^{2+} is heavily buffered, the density of sites required to regenerate and propagate the Ca^{2+} signal must be quite high. Have you any idea what that density is? How many of these elementary sites do you need in order to propagate a wave through the cell?

Berridge: That's an interesting question. In the case of muscle it might be possible to find that out, and there is some evidence for a high density of RYRs. But in other cell types we don't really have that information. It's quite hard to do immunohistochemistry on, say, InsP$_3$ receptors, although this is something we would like to know because the density of these sites has a marked effect in determining the rate of propagation. In fact, nature has done some beautiful experiments for us. For example, in the *Xenopus* oocyte the fertilization wave progresses faster at the animal pole than at the vegetal pole. This correlates with the density of InsP$_3$ receptors, which is much higher at the animal pole where the wave moves quickly.

Schöfl: I have a few questions regarding the localized Ca^{2+} puffs around the nucleus. Do they also occur at rest, which might be important for the control of Ca^{2+}-dependent genes expressed in the basal state? Are these Ca^{2+} puffs differentially regulated by distinct agonists? Is there any evidence that for example InsP$_3$-mobilizing agonists are better at enhancing these perinuclear changes in Ca^{2+} than agonists which predominantly activate Ca^{2+} influx through voltage-gated Ca^{2+} channels?

Berridge: With regard to the first point, we do see some activity at rest, but it is very low. We need to activate the cell to see these elementary events. It might just be fortuitous that we find these all around the nucleus because that's where most of the endoplasmic reticulum is located. The ER also spreads out into the periphery, but we don't see many puffs out there. It's a paradoxical situation in that the peripheral ER is located closer to the site of InsP$_3$ generation, yet when you start to activate the cell you see a lot of activity immediately around the nucleus. The Ca^{2+} released from one of these perinuclear puffs enters the nucleus very quickly and you can actually see it traversing through the nucleus and popping out at the other end.

Multiple pathways of ERK activation by G protein-coupled receptors

Thomas Gudermann

Institut für Pharmakologie und Toxikologie, Philipps-Universität Marburg, 35033 Marburg, Germany

Abstract. It has only recently been fully realized that G protein-coupled receptors and G proteins play crucial roles in the regulation of cell growth, differentiation and even tumour formation. Naturally occurring mutations in G protein-coupled receptors and in G protein α subunits result in uncontrolled cellular proliferation resulting in distinct human diseases. One important mechanism to transduce mitogenic signals from the cell membrane to the cell nucleus is the engagement of the extracellular signal-regulated kinase (ERK)–mitogen-activated protein kinase (MAPK) cascade. A multitude of distinct signal transduction pathways have been deciphered which connect G proteins with the ERK cascade. Both receptor and non-receptor tyrosine kinases play pivotal roles in these signalling pathways. Mitogenic signalling by G protein-coupled receptors can be regarded as a complex interplay between signals emanating from different classes of cell surface receptors which ultimately converge upon a small subset of central signalling proteins in the cell. The characterization of receptor-, G protein- and tyrosine kinase-specific contributions to mitogenic signalling in a particular cell and the identification of proteins serving as a point of convergence in the mitogenic signalling cascade may ultimately allow the design of novel pharmacological approaches to treat diseases involving unrestricted cell proliferation.

2001 Complexity in biological information processing. Wiley, Chichester (Novartis Foundation Symposium 239) p 68–84

Cellular differentiation and proliferation programs are determined by transcription factors, some of which are controlled by the extracellular signal-regulated kinase (ERK) subfamily of mitogen-activated protein kinases (MAPKs). One of the most extensively studied signalling cascades is activated by ligand-engaged receptor tyrosine kinases (RTKs) which recruit guanine nucleotide exchange factors for the monomeric GTPase Ras subsequent to RTK autophosphorylation and tyrosine phosphorylation of adaptor proteins like SHC and Grb2. Activated Ras subsequently engages the ERK–MAPK cascade involving the serine/threonine kinase Raf, MEK and finally ERKs (Fig. 1).

In addition to classical growth factors such as epidermal growth factor (EGF) and platelet-derived growth factor (PDGF), agonists acting at G protein-coupled

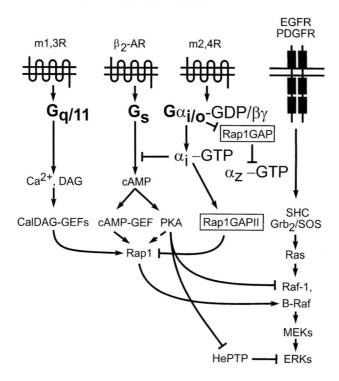

FIG. 1.   Rap1-mediated ERK activation. Signals emanating from G$_s$-coupled receptors feed into the ERK cascade at the level of Raf. In Raf-1 expressing cells, Gα_i-mediated inhibition of adenylyl cyclase may lead to derepression of PKA-mediated Raf inhibition. Additionally, cAMP directly activates guanine nucleotide exchange factors (GEFs) for Rap1. Agonist-bound G$_{q/11}$-coupled receptors initiate a rise in [Ca^{2+}]$_i$ and DAG, thereby directly activating CalDAG-GEFs for Rap1. GTP-loaded Gα_i activates a GTPase-activating protein (Rap1-GAPII) for Rap1 resulting in decreased Rap1 activity. As Rap1 is thought to compete with Raf for Ras binding, decreased Rap1 activity promotes signalling through the Ras/Raf-1/MEK/ERK pathway. In neuronal cells, GDP-loaded Gα_o may sequester Rap1-GAP and thus enhance signalling through the Rap1/B-Raf/MEK/ERK pathway. Rap1-GAP also interacts with GDP-liganded Gα_z and inhibits G$_z$-dependent signalling pathways. PKA-mediated phosphorylation of a haematopoietic phosphotyrosine phosphatase (HePTP) impairs the enzyme's ability to dephosphorylate ERKs. Solid lines indicate proven relations between signalling components; dashed lines represent putative relationships or multiple-step signalling between different components. β_2-AR, β_2-adrenergic receptor; CalDAG-GEF, calcium/DAG-stimulatable GEF; cAMP-GEF, cAMP-activated GEF; EGFR, epidermal growth factor receptor; ERK, extracellular signal-regulated kinase; GAP, GTPase activating protein; Grb2/SOS, complex of the adaptor protein growth-factor-receptor-bound protein 2 and the guanine nucleotide exchange factor son-of-sevenless; HePTP, haematopoietic phosphotyrosine phosphatase; MEK, mitogen-activated protein kinase/extracellular signal-regulated kinase kinase; m1,3R, m1 and m3 muscarinic acetylcholine receptor; m2,4R, m2 and m4 muscarinic acetylcholine receptor; PDGFR, platelet-derived growth factor receptor; PKA, cAMP-dependent protein kinase.

receptors (GPCRs) also play a role in differentiation and cellular transformation, and there is compelling evidence to suggest that GTPase-deficient G protein α subunits can behave as oncogenes. This paper presents a synopsis of the multiple signalling pathways leading to ERK activation by GPCRs and heterotrimeric G proteins. For more specialized information the reader is referred to several recent reviews (Gutkind 1998, Farfel et al 1999, Gudermann et al 2000).

Effects of G_s and cAMP on ERK activity

In NIH 3T3 cells, an increase in intracellular cAMP and activation of protein kinase A (PKA) reverses the oncogenic phenotype induced by constitutively active Ras. The reversal of transformation by cAMP is due to PKA-mediated inhibition of Raf kinase activity. In PC12 cells, however, cAMP does not antagonize the activation of ERKs by growth factors, but rather activates these kinases. The latter phenomenon requires the expression of B-Raf which in contrast to Raf-1 is not inhibited by PKA. The small GTPase Rap1 appears to be a selective activator of B-Raf by a mechanism analogous to Raf-1 activation by Ras (Fig. 1).

Three common second messengers, Ca^{2+}, diacylglycerol (DAG) and cAMP, are capable of activating Rap1 in a cell type-specific manner. Interestingly, phorbol ester-induced Rap1 activation is insensitive to inhibitors of PKC, and cAMP-mediated Rap activation does not require PKA. Recently, cAMP-binding proteins have been identified that have the ability to directly activate Rap1 (Kawasaki et al 1998, de Rooij et al 1998). These proteins are characterized by both cAMP-binding and guanine nucleotide exchange factor (GEF) domains and selectively activate Rap1 in a cAMP-dependent, but PKA-independent manner (see Fig. 1). In addition, a family of three Ca^{2+} and DAG-binding proteins (CalDAG-GEFI, II, and III) predominantly expressed in the central nervous system has recently been identified (Yamashita et al 2000). These proteins display GEF activity upon Ca^{2+} and DAG mobilization without requiring calmodulin and protein kinase C (PKC). CalDAG-GEFI is a GEF for Rap1 and R-Ras, while CalDAG-GEFII (identical to Ras-GRP) facilitates guanine nucleotide exchange in Ras and R-Ras (see Fig. 1). CalDAG-GEFIII has the broadest spectrum of Ras family GTPase substrates (H-Ras, R-Ras, Rap1) and affects differentiation and proliferation in a cell type-specific manner.

In lymphocytes a novel cross-talk mechanism between cAMP-dependent protein kinase and ERKs is realized by the haematopoietic protein tyrosine phosphatase HePTP, a negative regulator of ERK and p38 MAP kinases (see Fig. 1). HePTP dephosphorylates a critical tyrosine residue in the kinase activation loop. cAMP-dependent protein kinase phosphorylates HePTP in the kinase interaction motif, resulting in a decreased affinity of HePTP to its kinase substrates, the release of activated MAP kinases and transcription induction from the c-*fos* promoter (Saxena et al 1999).

ERK activation via $G\alpha_{i/o}$ subunits

Expression of GTPase-deficient $G\alpha_{i2}$ in Rat 1a cells results in constitutive activation of the MEK–ERK cascade by a poorly understood mechanism. Mochizuki et al (1999) showed that activated GTP-bound $G\alpha_i$ forms a complex with an N-terminally extended isoform of Rap1 GTPase activating protein (Rap1-GAPII), thus activating Rap1-GAPII and decreasing the cellular content of active, GTP-bound Rap1 (see Fig. 1). Considering the potential role of Rap1 as an antagonist of Ras signalling by scavenging Ras effectors like Raf1 (Bos 1998), a decrease in GTP-bound Rap1 is expected to amplify Ras signalling, and in HEK 293T cells ERK–MAPKs are activated by Rap1-GAPII overexpression independently of Ras activation (Mochizuki et al 1999).

A variation of the same general theme is illustrated by the modulation of Rap1 activity by $G\alpha_o$. Jordan et al (1999) observed that GDP-bound $G\alpha_o$ interacts with Rap1-GAP, thereby sequestering the exchange factor and raising the cellular amount of activated Rap1 (see Fig. 1). In PC12 cells, overexpression of unactivated $G\alpha_o$ resulted in ERK activation in a Rap1-GAP-dependent manner (Jordan et al 1999). Thus in neuronal cells, activation of a $G_{i/o}$-coupled receptor may antagonize cAMP-dependent signalling to ERK by lowering cellular cAMP levels on the one hand and by mitigating the Rap1-GTP/B-Raf/MEK/ERK cascade via release of $G\alpha_o$-sequestered Rap1-GAP. $G\alpha_z$, a member of the $G\alpha_i$ family highly expressed in brain, specifically interacts with Rap1-GAP in its activated form (Meng et al 1999). While this protein–protein interaction has no effect on the GAP activity of Rap1-GAP towards Rap1, it may modulate Rap1 signalling processes by relocalizing Rap1-GAP within the cell. Conversely, the interaction between $G\alpha_z$ and Rap1-GAP sequesters the $G\alpha$ subunit and prevents interactions with RGS proteins or effectors such as adenylyl cyclase. Thus, Rap1-GAP may act as an integrator between G_z and Rap1 signal transduction pathways.

ERK activation via $G\beta\gamma$ subunits released from activated G_i proteins

Mitogenic signals mediated by G_i-coupled receptors are thought to be chiefly transmitted via $G\beta\gamma$ subunits (Gutkind 1998). There is convincing evidence to support a role of the Src family of tyrosine kinases in mitogenic signalling via G_i-coupled receptors. For the COS-7 cell model it has been postulated that Src is responsible for GPCR-mediated SHC phosphorylation and for the use of the adaptor protein as a point of entry into the Ras/ERK signalling cascade (Fig. 2). Other signalling molecules implicated in linking $G\beta\gamma$ to the ERK cascade are the

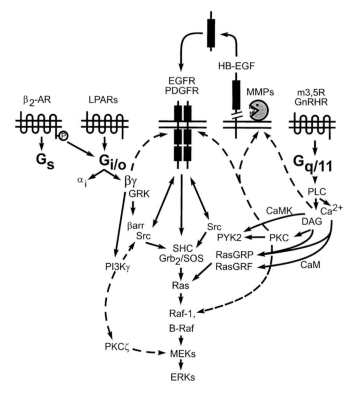

FIG. 2. Activation of ERK-MAP kinases by $G_{i/o}$- and $G_{q/11}$-coupled receptors. While G protein $\beta\gamma$ subunits play a central role in ERK activation via G_i-coupled receptors, products of phospholipase C activity such as DAG and Ca^{2+}, and protein kinase C isoforms activated by these products are key signalling intermediates linking $G_{q/11}$-coupled receptors with the ERK-MAPK cascade. Increases in $[Ca^{2+}]_i$ and in PKC activity are suggested to activate matrix metallo-proteinases (MMPs) at the cell surface. Activated MMPs may release heparin-bound EGF-like growth factors (HB-EGF) which initiate a classical RTK signalling cascade. Src-like kinases are recruited by β-arrestins bound to GRK-phosphorylated G-protein-coupled receptors or by PYK2, a cytosolic tyrosine kinase related to focal adhesion kinase. As yet, it is not clear whether Src kinases play a role upstream or downstream of activated receptor tyrosine kinases such as EGF or PDGF receptors. See text for further details. β_2-AR, β_2-adrenergic receptor; βarr, β-arrestin; CaM, calmodulin; CaMK, CaM protein kinase; DAG, diacylglycerol; EGFR, epidermal growth factor receptor; ERK, extracellular signal-regulated kinase; GnRHR, gonadotropin-releasing hormone receptor; GEF, guanine nucleotide exchange factor; GnRH, gonadotropin-releasing hormone; Grb2/SOS, complex of the adaptor protein growth-factor-receptor-bound protein 2 and the guanine nucleotide exchange factor son-of-sevenless; GRK, G-protein-coupled receptor kinase; HB-EGF, heparin-bound EGF-like growth factor; LPA, lysophosphatidic acid; LPAR, LPA receptor; m3,5R, m3 and m5 muscarinic acetylcholine receptor; MEK, mitogen-activated protein kinase/extracellular signal-regulated kinase kinase; MMP, matrix metalloproteinase; PDGFR, platelet-derived growth factor receptor; PI3K, phosphatidylinositol 3-kinase; PKC, protein kinase C; PLC, phospholipase C; PYK2, proline-rich tyrosine kinase 2; RasGRF, RasGRP, guanine nucleotide exchange factors for Ras.

protein tyrosine phosphatase PTP1C and Ras guanine nucleotide-releasing factor (Ras-GRF) (Cdc25^{Mm}), a Ras-GEF highly enriched in neuronal cells (for review see Gudermann et al 2000). Ras-GRF also exhibits Rac1-specific GEF activity when activated by G$\beta\gamma$-mediated signals. Phosphorylation of Ras-GRF by the non-receptor tyrosine kinase Src is sufficient for the induction of GEF activity towards Rac, indicating that Src may be located downstream of G$\beta\gamma$ to modulate Ras-GRF substrate specificity (Kiyono et al 2000).

ERK activation via G$_{q/11}$-coupled receptors

PKC isotypes are central signalling molecules coupling the G$_{q/11}$ family of G proteins to ERKs. Activation of PKC by phorbol esters results in robust ERK activation in most cell types by a mechanism that is still unclear. One potential mechanism to explain ERK activation by PKC is derived from the observation that PKCα directly phosphorylates and activates Raf-1 *in vitro* and in an NIH 3T3 cell clone (reviewed in Gutkind 1998) (see Fig. 2). However, it was recently shown that Raf must be associated with Ras-GTP for its activation by PKC (Marais et al 1998), so that the primary role of PKC for Raf activation appears to be the activation of Ras, which would provide for membrane anchoring of Raf. Phorbol esters were demonstrated to stimulate Ras-GTP accumulation, but the activation of Raf by PKC was not blocked by N17Ras (Marais et al 1998). N17Ras, which sequesters Ras-GEFs, for example SOS, abrogates Raf activation by RTKs, suggesting that PKC activates Ras by a mechanism distinct from that employed by RTKs, probably via a novel GEF distinct from SOS.

Ca^{2+}- and cytoskeleton-dependent ERK activation

In PC12 cells, Ca^{2+} transients alone were reported to be sufficient to trigger ERK activation. Special emphasis has been laid on the role of proline-rich tyrosine kinase 2 (PYK2, also known as cell adhesion kinase β [CAKβ], related adhesion focal tyrosine kinase [RAFTK], Ca^{2+} dependent protein tyrosine kinase [CADTK]), a non-receptor tyrosine kinase related to focal adhesion kinases, which appears to be involved in ERK activation by G$_{q/11}$-coupled receptors via a direct interaction of the tyrosine phosphorylated PYK2 with c-Src (see Schlaepfer et al 1999). PYK2 expressed at high levels in cells of neuronal origin and in cells of the haematopoietic lineage, is independently activated by elevations of [Ca^{2+}]$_i$ and by phorbol esters (see Fig. 2). The activated tyrosine kinase subsequently mediates Ras-dependent ERK activation via interaction with Src, tyrosine phosphorylation of Shc and Shc–Grb2/SOS complex formation. These observations underpin a G$_q$- and

Ca^{2+}-mediated signalling pathway that utilizes the same intermediates as $G\beta\gamma$- and RTK-mediated pathways.

Ligand-independent activation of receptor tyrosine kinases

The convergence of signalling pathways at the level of or even upstream of Shc proteins highlighted a previously unrecognized degree of cross-talk between the GPCR and RTK signalling pathways. In each case tyrosine kinase activity is required for mitogenic signalling. The identity of the tyrosine kinases involved in GPCR signalling is still a matter of debate, but there is mounting evidence that RTKs are of central importance to GPCR signalling (Zwick et al 1999a). Ligand-independent activation (transactivation) of the EGF receptor (EGFR) appears to be a general phenomenon evoked by various $G_{q/11}$- or G_i-coupled receptors in different cellular settings. Later on the concept evolved that cross-talk between different classes of cell surface receptors, i.e. GPCRs and RTKs, is a general feature, because the GPCR ligand pysophosphatidic acid (LPA) induces ligand-independent tyrosine phosphorylation of the EGFR or of the PDGFβ receptor depending on the cellular setting (Herrlich et al 1998).

The molecular mechanism of RTK activation via GPCRs is not thoroughly understood at present. Although several studies analysing conditioned cell culture media failed to obtain evidence for a release of growth factors which would activate their respective receptors subsequent to GPCR stimulation, such a mechanism is hard to rule out. In CHO cells, shedding of membrane-anchored heparin-binding EGF-like growth factor (HB-EGF) can be effectively induced by Ca^{2+} ionophore and phorbol ester treatment (see Fig. 2). In several cell systems the shedding process executed by matrix metalloproteinases appears to be positively regulated by MAPKs (ERK, p38) (Fan & Derynck 1999). Time course analyses indicate that 12-O-tetradecanoyl-13-phorbol aceteate (TPA)-induced ERK activation occurs within 5 min and precedes soluble heparin-binding EGF-like growth factor (HB-EGF) release which can be observed after 10–20 min (Gechtman et al 1999). These findings can be interpreted to mean that Ca^{2+}- and TPA-induced ERK activation is located upstream of HB-EGF shedding.

Despite the latter kinetic reasoning a sound body of evidence underpins a significant contribution of metalloproteinase-mediated cleavage of proHB-EGF to GPCR-dependent EGFR transactivation (Prenzel et al 1999). Inhibition of proHB-EGF cleavage was shown to preclude GPCR-dependent EGFR activation, and proHB-EGF shedding was noted 10 min after LPA or TPA addition to COS cells. Preincubation of PC3 prostate cancer cells with batimastat, a potent non-selective inhibitor of metalloproteinases, abolished EGFR tyrosine phosphorylation upon neuropeptide or TPA challenge (Prenzel et al 1999). Although the identity of the metalloproteinase involved still remains elusive, these studies favour a transactivation mode which essentially reflects the classical way of RTK activation by growth factors and highlights a role for membranous proteinases as therapeutic targets.

In vascular smooth muscle cells as well as in neuronal PC12 cells a rise in $[Ca^{2+}]_i$ appears to be sufficient to trigger growth factor-independent tyrosine phosphorylation of RTKs (Zwick et al 1999a, Gudermann et al 2000). In these cellular settings Ca^{2+}-activated tyrosine kinases of the PYK2 family have recently been placed upstream of RTKs in the signalling pathway from $G_{q/11}$-coupled receptors to ERKs (reviewed in Gudermann et al 2000). In T84 intestinal epithelial cells carbachol-initiated EGFR tyrosine phosphorylation is brought about by a signalling pathway involving increases in intracellular Ca^{2+}, calmodulin, PYK2 and Src (Keely et al 2000), and PYK2 can be co-immunoprecipitated with EGFR in a Ca^{2+}-dependent manner. In breast cancer cells, the association of PYK2 and ErbB2 appears to be indirect and mediated by Src (Zrihan-Licht et al 2000). Although EGFR transactivation and PYK2 tyrosine phosphorylation have recently been described as two distinct and unrelated Ca^{2+}-dependent events in PC12 cells (Zwick et al 1999b), it is presently not possible to exclude a role of non-receptor tyrosine kinases PYK2 and Src in triggering RTK transactivation. Alternatively, experiments on the UV response indicated that inactivation of a phosphotyrosine phosphatase (PTP) is critically involved in RTK activation (Gross et al 1999), a process that is attributed to the UV-induced generation of reactive oxygen intermediates.

Interplay between different classes of cell surface receptors appears to be a general principle, because RTKs do not only cross-talk with GPCRs but also with cell adhesion molecules such as integrins (Boudreau & Jones 1999). RTKs can additionally be activated by cytokine action. In LNCaP prostate carcinoma cells IL-6 induces tyrosine phosphorylation of ErbB2 and ErbB3 and association of the IL-6 receptor gp130 subunit with ErbB2 (reviewed in Zwick et al 1999a). ErbB2 tyrosine kinase activity is a prerequisite for IL-6 signalling to ERK, and thus the ErbB2-specific tyrphostin AG879 precludes IL-6-induced MAPK activation. In contrast to GPCR- or IL-6 receptor-mediated RTK transactivation which necessitates intrinsic RTK tyrosine kinase activity for downstream signalling, the EGFR is directly phosphorylated by the cytosolic tyrosine kinase Jak2 following growth hormone stimulation (reviewed in Zwick et al 1999a).

Collectively, these findings place RTKs at a central position in many signal transduction pathways enabling them to modulate and integrate various extracellular stimuli. It is likely, however, that transactivation of RTKs is a principally dispensable event for GPCR-induced signalling to the nucleus and that RTKs are utilized to shape the kinetics of signal transmission to the nucleus which would nevertheless occur without RTK contribution (Grosse et al 2000a). There is recent evidence that the population of RTKs activated by growth factors is distinct from that transactivated via GPCRs, implicating a unique activation pathway separable from the one engaged by endogenous tyrosine kinase ligands (Heeneman et al 2000). Many groups are currently dedicating considerable effort to unravel intricate signalling networks and to identify relevant, cell-specific pathways.

Contribution of the receptor internalization machinery to ERK activation

In several cell systems signal transmission from GPCRs to ERKs is ablated subsequent to inhibition of clathrin-mediated endocytosis. Furthermore, it was shown that Src-mediated tyrosine phosphorylation of dynamin involved in fission of the budding clathrin-coated vesicle from the plasma membrane is a requirement for GPCR internalization and ERK signalling (Luttrell et al 1999). These findings have been interpreted to suggest that agonist-binding to the β_2-adrenergic receptor results in β-arrestin-dependent formation of a signalling complex consisting of the heptahelical receptor, β-arrestin and Src, which would then, through receptor internalization and Src-mediated Shc tyrosine phosphorylation, engage the Ras/ERK signalling cascade (Luttrell et al 1999) (see Fig. 2). By interacting with the catalytic domain of Src, β-arrestin recruits tyrosine kinase activity to the cell membrane to enable phosphorylation of key components of the endocytotic machinery such as dynamin and clathrin heavy chains (Wilde et al 1999, Miller et al 2000). This overall model of GPCR-dependent ERK activation circumvents the need for second messenger production, because the active receptor conformation is the decisive trigger for signalling. The role of $\beta\gamma$ subunits would primarily be the recruitment of receptor kinases to the agonist-bound receptor to secure phosphorylation of cytoplasmic receptor domains required for the interaction with β-arrestin.

However, GPCR internalization via clathrin-coated pits does not appear to be a universal prerequisite for ERK activation by G_i-coupling receptors, because the α_2-adrenergic as well as opioid receptors are able to engage the ERK–MAPK cascade under conditions which preclude receptor internalization (reviewed in Gudermann et al 2000). Mammalian exclusively $G_{q/11}$-coupled gonadotropin-releasing hormone (GnRH) receptors (Grosse et al 2000b) completely lack a cytoplasmic C-terminal tail and are regulated in a β-arrestin-independent fashion. None the less, GnRH challenge results in a rapid GTP-loading of Ras and increased ERK activity (Grosse et al 2000a), thus challenging the concept that GPCR endocytosis is generally required for ERK activation. In a systematic, comparative study DeFea et al (2000) demonstrated that both the wild-type PAR2 receptor and also a PAR2 receptor mutant defective in β-arrestin binding are able to activate ERKs, yet by distinct pathways. ERK activation by wild-type

PAR2 is accompanied by the formation of a multi-protein signalling complex consisting of the internalized receptor, β-arrestin, Raf-1 and activated ERKs, the latter being confined to the cytosol and devoid of mitogenic potential. The above-mentioned PAR2 receptor mutant was able to mediate ERK activation without concomitant signalling complex formation. Activated ERKs were now allowed access to the nucleus, resulting in increased cell proliferation (DeFea et al 2000). Thus, while GPCR-mediated ERK activation does not require GPCR internalization, the formation of an internalized signalling complex may ensure appropriate subcellular localization of ERK activity and thereby determine the mitogenic potency of a given agonist.

Conclusions concerning the importance of receptor endocytosis for GPCR-mediated ERK activation have been drawn from experiments employing a dynamin mutant (K44A) defective in GTP binding and hydrolysis. In cells overexpressing this mutant dynamin, coated pits fail to become constricted and to bud. While in many cell systems clathrin-mediated endocytosis is required for GPCR-dependent ERK activation, the dynamin mutant does not allow us to address the question of whether it is GPCR endocytosis or rather internalization of other signalling proteins such as activated MEK (Kranenburg et al 1999) which represents the crucial step in the ERK activation cascade. Clathrin-mediated endocytosis and downstream receptor signalling is not only required for GPCR internalization, but also for EGFR endocytosis, and Src-mediated tyrosine phosphorylation of clathrin heavy chains is a crucial prerequisite for effective EGFR internalization (Wilde et al 1999). A comparison of ERK activation cascades initiated by sequestering versus non-sequestering GPCRs provided evidence for the concept that RTKs or other downstream signalling proteins rather than GPCRs have to engage the endocytotic pathway (Pierce et al 2000). In the case of the β_2-adrenoceptor, isoproterenol-induced ERK activation is realized by the assembly of a multireceptor complex made up of the β_2-adrenoceptor, EGFR and interaction of the transactivated RTK with components of the endocytotic machinery (Maudsley et al 2000).

Conclusions

G protein-mediated signal transduction is realized as a complex signalling network with converging and diverging transduction steps at every coupling interface (Gudermann et al 1996). A new level of complexity has been added by realizing that cross-talk exists between different classes of cell surface receptors (i.e. GPCRs, RTKs, cytokine receptors, extracellular matrix receptors) at the level of the receptors and further downstream between the signalling cascades. The delineation of relevant signalling pathways shaping the intricate cross-talk between different classes of cell surface receptors may ultimately help develop novel pharmacological intervention strategies.

Acknowledgements

The author's own work reported herein was funded by the Deutsche Forschungsgemeinschaft. Whenever possible, review articles rather than original reports were listed. The author apologizes to all researchers whose work could not be cited due to strict space limitations.

References

Bos JL 1998 All in the family? New insights and questions regarding interconnectivity of Ras, Rap1 and Ral. EMBO J 17:6776–6782

Boudreau N J, Jones PL 1999 Extracellular matrix and integrin signalling: the shape of things to come. Biochem J 339:481–488

de Rooij J, Zwartkruis F J, Verheijen MH et al 1998 Epac is a Rap1 guanine-nucleotide-exchange factor directly activated by cyclic AMP. Nature 396:474–477

DeFea KA, Zalevsky J, Thoma MS, Déry O, Mullins RD, Bunnett NW 2000 β-arrestin-dependent endocytosis of proteinase-activated receptor 2 is required for intracellular targeting of activated ERK1/2. J Cell Biol 148:1267–1281

Fan H, Derynck R 1999 Ectodomain shedding of TGF-α and other transmembrane proteins is induced by receptor tyrosine kinase activation and MAP kinase signaling cascades. EMBO J 18:6962–6972

Farfel Z, Bourne HR, Iiri T 1999 The expanding spectrum of G protein diseases. N Engl J Med 340:1012–1020

Gechtman Z, Alonso JL, Raab G, Ingber DE, Klagsbrun M 1999 The shedding of membrane-anchored heparin-binding epidermal-like growth factor is regulated by the Raf/mitogen-activated protein kinase cascade and by cell adhesion and spreading. J Biol Chem 274:28828–28835

Gross S, Knebel A, Tenev T et al 1999 Inactivation of protein-tyrosine phosphatases as mechanisms of UV-induced signal transduction. J Biol Chem 274:26378–26386

Grosse R, Roelle S, Herrlich A, Höhn J, Gudermann T 2000a Epidermal growth factor receptor tyrosine kinase mediates Ras activation by gonadotropin releasing hormone. J Biol Chem 275:12251–12260

Grosse R, Schmid A, Schöneberg T et al 2000b Gonadotropin-releasing hormone receptor initiates multiple signaling pathways by selectively coupling to $G_{q/11}$ proteins. J Biol Chem 275:9193–9200

Gudermann T, Kalkbrenner F, Schultz G 1996 Diversity and selectivity of receptor–G protein interaction. Annu Rev Pharmacol Toxicol 36:429–459

Gudermann T, Grosse R, Schultz G 2000 Contribution of receptor/G protein signaling to cell growth and transformation. Naunyn Schmiedebergs Arch Pharmacol 361:345–362

Gutkind JS 1998 Cell growth control by G protein-coupled receptors: from signal transduction to signal integration. Oncogene 17:1331–1342

Heeneman S, Haendler J, Saito Y, Ishida M, Berk BC 2000 Angiotensin II induces transactivation of two different populations of the PDGFβ-receptor: Key role for the adaptor protein Shc. J Biol Chem 275:15926–15932

Herrlich A, Daub H, Knebel A et al 1998 Ligand-independent activation of platelet-derived growth factor receptor is a necessary intermediate in lysophosphatidic acid-stimulated mitogenic activity in L cells. Proc Natl Acad Sci USA 95:8985–8990

Jordan JD, Carey KD, Stork PJS, Iyengar R 1999 Modulation of Rap activity by direct interaction of $G\alpha_o$ with Rap1 GTPase-activating protein. J Biol Chem 274:21507–21510

Kawasaki H, Springett GM, Mochizuki N et al 1998 A family of cAMP-binding proteins that directly activate Rap1. Science 282:2275–2279

Keely SJ, Calandrella SO, Barrett KE 2000 Carbachol-stimulated transactivation of epidermal growth factor receptor and mitogen-activated protein kinase in T_{84} cells is mediated by intracellular Ca^{2+}, PYK-2, and p60[src]. J Biol Chem 275:12619–12625

Kiyono M, Kaziro Y, Satoh T 2000 Induction of Rac-guanine nucleotide exchange activity of Ras-GRF1/CDC25[Mm] following phosphorylation by the nonreceptor tyrosine kinase Src. J Biol Chem 275:5441–5446

Kranenburg O, Verlaan I, Moolenaar WH 1999 Dynamin is required for the activation of mitogen-activated protein (MAP) kinase by MAP kinase kinase. J Biol Chem 274:35301–35304

Luttrell LM, Daaka Y, Lefkowitz RJ 1999 Regulation of tyrosine kinase cascades by G-protein-coupled receptors. Curr Opin Cell Biol 11:177–183

Marais R, Light Y, Mason C, Paterson H, Olson MF, Marshall CJ 1998 Requirement of Ras-GTP-Raf complexes for activation of Raf-1 by protein kinase C. Science 280:109–112

Maudsley S, Pierce KL, Zamah AM et al 2000 The β_2-adrenergic receptor mediates extracellular signal-regulated kinase activation via assembly of a multi-receptor complex with the epidermal growth factor receptor. J Biol Chem 275:9572–9580

Meng J, Glick JL, Polakis P, Casey PJ 1999 Functional interaction between $G\alpha_z$ and Rap1GAP suggests a novel form of cellular cross-talk. J Biol Chem 274:36663–36669

Miller WE, Maudsley S, Ahn S, Khan KD, Luttrell LM, Lefkowitz RJ 2000 β-arrestin1 interacts with the catalytic domain of the tyrosine kinase c-SRC. Role of β-arrestin1-dependent targeting of c-SRC in receptor endocytosis. J Biol Chem 275:11312–11319

Mochizuki N, Ohba Y, Kiyokawa E et al 1999 Activation of ERK/MAPK pathway by an isoform of rap1GAP associated with $G\alpha_i$. Nature 400:891–894

Pierce KL, Maudsley S, Daaka Y, Luttrell LM, Lefkowitz RJ 2000 Role of endocytosis in the activation of the extracellular signal-regulated kinase cascade by sequestering and nonsequestering G protein-coupled receptors. Proc Natl Acad Sci USA 97:1489–1494

Prenzel N, Zwick E, Daub H et al 1999 EGF receptor transactivation by G-protein-coupled receptors requires metalloproteinase cleavage of proHB-EGF. Nature 402:884–888

Saxena M, Williams S, Taskén K, Mustellin T 1999 Crosstalk between cAMP-dependent kinase and MAP kinase through a protein tyrosine phosphatase. Nat Cell Biol 1:305–311

Schlaepfer DD, Hauck CR, Sieg DJ 1999 Signaling through focal adhesion kinase. Prog Biophys Mol Biol 71:435–478

Wilde A, Beattie EC, Lem L et al 1999 EGF receptor signaling stimulates SRC kinase phosphorylation of clathrin, influencing clathrin redistribution and EGF uptake. Cell 96:677–687

Yamashita S, Mochizuki N, Ohba Y et al 2000 CalDAG-GEFIII activation of Ras, R-Ras, and Rap1. J Biol Chem 275:25488–25493

Zrihan-Licht S, Fu Y, Settleman J et al 2000 RAFTK/Pyk2 tyrosine kinase mediates the association of p190 RhoGAP with RasGAP and is involved in breast cancer cell invasion. Oncogene 19:1318–1328

Zwick E, Hackel PO, Prenzel N, Ullrich A 1999a The EGF receptor as central transducer of heterologous signalling systems. Trends Pharmacol Sci 20:408–412

Zwick E, Wallasch C, Daub H, Ullrich A 1999b Distinct calcium-dependent pathways of epidermal growth factor receptor transactivation and PYK2 tyrosine phosphorylation in PC12 cells. J Biol Chem 274:20989–20996

DISCUSSION

Dolmetsch: Do you know what the Ca^{2+} sensor for PIP2 is? I have often wondered what the Ca^{2+} sensor for the MAP kinase cascade is. This cascade is clearly Ca^{2+} activated, yet PIP2 itself does not bind Ca^{2+}.

Gudermann: The simple answer is that it is not known. The only observation is that an elevation of Ca^{2+} is sufficient to lead to ERK activation. There are some ideas that perhaps CaM kinases are involved here, so this could be a link between Ca^{2+} and PIP2.

Dolmetsch: The problem with the ERK cascade is the problem shared by all signalling cascades: there are clearly lots of players and they all interact with each other. If you do a literature search for any one of these pathways you get many thousands of papers, half saying one thing and the other half saying things that are apparently contradictory. The current approach for unravelling this system is for everyone to work on one small story. The typical postdoc or graduate student will work on the activation of some target by ERK or PKB or whatever. The problem with this approach is that it produces lots of contradictory results. Do you have any idea of how you could integrate this into something that is reliable?

Gudermann: One has to ask specific questions and explore them in depth in the cell. Perhaps there is no way around these little stories. We were interested in understanding how GnRH signalling works in the gonadotrope, so we concentrated on gonadotropic pituitary cells and tried to assemble the signalling pathway in this specific cell, knowing that if we took the same signalling components in a slightly different cell line, things would be completely different. I don't know what you mean by 'making sense' out of this: is there really one universal concept in cell signalling? I am not sure there is.

Sejnowski: Is the source of apparent unreliability the fact that people are using different cells or different ligands? Are they doing exactly the same experiments and getting different results?

Dolmetsch: Of course not. The difference is partly due to different cells and different techniques. Part of it, however, is the way in which signalling in general is studied. Typically, when people attempt to characterize an interaction in a signalling pathway, they use lots of different cell lines. Neurons are hard to transfect and it is difficult to do biochemistry on neurons, so half of the experiments are done in 293 cells. The problem here is that in the end you have these papers that tell stories which in fact are a mish mash of what is happening in a lot of different cells.

Gudermann: We have to go back and do the hard job of studying those cells that endogenously express the receptors and signalling components. If we really want to understand whether this has any physiological meaning, we have to go as close to physiology as we can.

Kahn: The problem isn't the reliability of the data. For the most part, the data are the data: people haven't done bad experiments; they have done what they said they did and have reported the results of these. I think part of the confusion comes from the stoichiometry of molecules when a reconstituted cell is made. I think it is very important to look in physiologically relevant cells at physiological concentrations of these proteins. Along these lines, with regard to the GnRH story and the EGF receptor role, is the role of the ERK signalling in GnRH action ultimately via the release of gonadotropins? GnRH may have multiple effects on that cell and the EFK pathway may just be involved in some of these.

Gudermann: We don't know yet. These are the questions that we are addressing right now. The first thing we did was to look at transcriptional events, and there are no obvious differences there. This does not mean that the accelerated time course of ERK activation via the EGF receptor has no role, but at the moment we have no idea what it does physiologically.

Kahn: I have been thinking about the idea of the signal from, say, a GPCR, coming to an intracellular domain, creating mediators which then come back and activate another membrane receptor to create a new intracellular signal. Is there are rationale for why the system might adopt this seemingly inefficient method of signal transduction?

Gudermann: There are a number of potential answers. One has to do with the strength of the signals. The signal goes into the ERK cascade after the binding of growth factor and the activation of a GPCR. Usually this is a much lower level than what is seen after the cell has been stimulated by EGF. Again, this does not necessarily mean that this signal has no physiological significance. The easiest explanation would be that such a *trans*-activated EGF receptor may only serve as a scaffold molecule for the assembly of components, so that the signalling pathway can progress into the cell in a coordinated fashion. But there is a clear difference in the strength of the signal. There are some papers looking at the contribution of colonic epithelial cells which claim that you need the input of a RTK to see a physiological effect. But studies addressing a physiological endpoint are very rare.

Sejnowski: There are many possible benefits of this sort of convoluted signal pathway, such as additional points of regulation, or the need for signal validation to make sure it is not just random fluctuation or system noise.

Gudermann: The idea that it is a validation or amplification system is an attractive hypothesis. But there are very few studies in which people have tried to eliminate the contribution of this amplifier-like signal system. There is some evidence that this kind of cross-talk between membrane receptors not only exists between receptor tyrosine kinases and GPCRs, but also between cell adhesion molecules and cytokine receptors.

Eichele: Coming back to the issue of physiological and pharmacological significance, could it be time to return to the organismal level with some of these questions? I realise that GnRH neurons can't be studied in *Caenorhabditis elegans* or *Drosophila*, but these model organisms have been wonderfully developed for addressing signalling in the organismal context. Even in the mouse we have, at least in principal, the tools to address these questions. We can get away from cells.

Iyengar: Some of us are still stuck at the cellular level. I'd like to share some of the experience from my own lab over the last 8–10 years. 10 years ago we published a paper stating that activated $G_\alpha o$ transforms NIH 3T3 cells. I ruined at least two graduate students by pushing them to find out how αo activated MAP kinase. It didn't work at all. It turns out that in those cells, for reasons I still don't fully understand, αo engages Src to activate STAT3. The entire pathway operates without touching MAP kinase in NIH 3T3 cells. We have repeated the same experiments that Bob Lefkowitz and others have done that have shown αo activation of MAP kinase in CHO cells, and they work for us as well. There are questions that remain unanswered at the cellular level. We don't understand what the context of these connections is, and until we do, we are effectively building on a second level of black boxes if we go on to the organismal level. Src shows up in so many contexts with respect to this type of signalling, in channel regulation, engagement of integrins and so on.

Sejnowski: There is a real danger that some of the phenomena being studied are basically tissue culture artefacts. This is true of primary cell lines of neurons; the transformed cell lines are even less normal.

Iyengar: We just have to be careful that we don't overexpress the molecules we are studying. These things can be controlled.

Kahn: The presumption there is that we know the correct concentration of the different components that are put back. I am not sure we do yet.

Eichele: Src is a good case: the Src knockout mouse doesn't show what it is supposed to show. I appreciate cellular studies, but I would like to see genetic experiments done at the organismal level first.

Berridge: If Src has been knocked out and there is no phenotype, what do we do then?

Pozzan: I think that cell lines deserve an advocate. My impression is that studies on cell lines indicate the existence of potential signaling pathways. Then we have to verify whether those potential pathways are actually exploited in the organism. Knockout mice very often don't show a phenotype or show an unexpected phenotype. In the case of the growth factors that you mentioned, another aspect that needs to be considered is the amount of time for which the receptor is engaged. For example, muscarinic receptors are potentially mitogenic because they can activate the MAP kinase pathway, but most often people use carbachol (that is not metabolized) rather than the physiological agonist acetylcholine, that is rapidly metabolized. In other words, in real life it is unlikely that acetylcholine will activate the muscarinic receptor for long enough to effectively turn on a mitogenic signal.

Gudermann: I agree. I would like to reemphasize the point that we should be aware of exactly which questions we are asking and what answers we are going to get. If one asks the question about what potential interactions are there, and if one only interprets them as potential interactions and does not draw any unjustified conclusions as to what is happening in real life, these sorts of studies are valuable.

Dolmetsch: Knockout animals are useful for answering specific sorts of questions. They are extremely useful for questions of development. But if you want to test the function of a particular molecule in an adult, you need a slightly different approach, such as using an inducible knockout. Just about every time I look at the journal *Neuron*, there are papers in which someone has produced a knockout mouse and tested in the Morris water maze, showing that it has a learning deficit, or that it learns better. These experiments are completely uninformative. They tell us virtually nothing about the function of the molecule that has been knocked out. I am not so sure that making a knockout is the way forward, if we don't really understand what that molecule does.

von Herrath: I have a point that has been increasingly worrying me over the last few years. If I take a field such as this, which is not my own, and do a literature search, this will pull up say 10 000 papers. The human mind is not made to integrate all this information and make sense of it. Can all this information, which is probably very useful, be put into some kind of computer-guided algorithm that would integrate it, together with all the experimental details contained in the papers? If we only look at the abstracts we will miss all the technical details. This sort of approach may provide a route to a better mechanistic understanding of some of these areas. Or is this approach complete nonsense at this stage?

Berridge: You may or may not be aware of the initiative Al Gilman has put forward, the Alliance for Cell Signalling. This is intended to do just what you have mentioned. Again, the issue about cloned cells comes up. I am very much against the use of cloned cells, mainly because they have undergone an ill-defined immortalization phenomenon, which makes them very different to primary cells. We really have to tackle primary cells, and this is what Al Gilman has decided. Therefore, the Alliance for Cell Signalling will study two cell types: the ventricular cardiac cell and B lymphocytes. The whole emphasis is on looking at primary cells.

Iyengar: That is just one point of view. I think there is a lot more to be learned from cloned cells.

Berridge: There is a lot more to be learned about the artefacts!

Iyengar: There are many people who think that there are other ways to do this research from the way you suggested. We have learned a lot from cloned cells. It would be disingenuous to suggest that we haven't made progress because we have been using these cell lines. All of cancer biology has come from studying transformed cell lines. Even last year when Weinberg published his paper on 293 cells, we still learned something new. I agree about the limitations, but I think we can still learn quite a bit more.

Sejnowski: I don't think that the suggestion is that we stop the old research, just that we initiate a new line that potentially will give a different picture and allow us to see something that we can't get from a transformed cell line.

Heterogeneity of second messenger levels in living cells

Manuela Zaccolo, Luisa Filippin, Paulo Magalhães and Tullio Pozzan[1]

Department of Biomedical Sciences, CNR Centre of Biomembranes, University of Padua, Viale G. Colombo 3, 35121 Padua, Italy

Abstract. Over the last years we have utilised chimeras from aequorin and green fluorescent protein (GFP) to monitor the dynamics of second messenger levels in living cells. In this contribution we address two problems, i.e. the complexity of Ca^{2+} handling by mitochondria and the localization of cAMP signalling. As to the first, we here demonstrate that physiological increases in mitochondrial Ca^{2+}, monitored with selectively localized recombinant aequorin, concern a sub-population of organelles that is stably and selectively associated with the endoplasmic reticulum. As to cAMP, we describe the use of a novel probe to monitor its changes in living cells, that takes advantage of the phenomenon of fluorescence resonance energy transfer (FRET) between suitable GFPs linked to the regulatory and catalytic subunits of protein kinase A (PKA). When cAMP is low the two fluorophores are in close proximity and generate FRET while increasing levels of cAMP determine progressive reduction of FRET as the two subunits (linked to the GFPs) diffuse apart. We also demonstrate that by using such cAMP sensor, localized increase of this second messenger can be observed upon selective stimulation of plasma membrane receptors.

2001 Complexity in biological information processing. Wiley, Chichester (Novartis Foundation Symposium 239) p 85–95

When a receptor binds its ligand on the outer surface of the plasma membrane, it undergoes a conformational change that permits the initiation of a cascade of events leading, eventually, to the activation of specific cellular responses. These membrane receptors can be subdivided into three major categories: those endowed with intrinsic catalytic activities (e.g. the growth factor receptors), those coupled to G proteins (e.g. muscarinic acetylcholine receptors), and those endowed with ion channel properties (e.g. the ionotropic glutamate receptors). Despite major differences in the mechanisms of signal transduction, multiple

---

[1]This paper was presented at the symposium by Tullio Pozzan to whom correspondence should be addressed.

communication points exist between these signalling pathways. In most cases — though not in all — the activation of the receptor results, directly or indirectly, in the alteration of the concentration of one or more intracellular second messengers. These are small, soluble molecules — the concentration of which can vary rapidly and transiently within the cell — and their binding to specific effector proteins controls a variety of cellular phenomena. To date, although hundreds of receptors are known to be expressed in different cell types, the number of known second messengers is extremely small. The classical second messengers are Ca^{2+}, inositol-1,4,5-trisphosphate ($InsP_3$), diacylglycerol, cAMP and cGMP. A few other compounds (NO, phosphatidylinositol-3,4,5-trisphosphate [PIP_3], cADP ribose, NAADP, and arachidonic acid) could be included in this class of molecules, though they can hardly be defined as second messengers *sensu stricto*. Amongst the *bona fide* second messengers, it is noteworthy that the only known action of $InsP_3$ is that of inducing Ca^{2+} mobilization from stores — all its biological effects are thus mediated through Ca^{2+} ions. The same reasoning applies to cADP ribose and NAADP. Succinctly, the many different receptors that are involved in the control of an ample variety of biological processes (from contraction to secretion, from phagocytosis to cell division), exert their biological action through the modulation of intracellular levels of very few messengers. In other words, many receptors share the same second messenger yet promote different cell responses. How can distinct cellular functions be controlled by the same signal?

Different strategies can be envisaged to achieve this purpose, the simplest being cell differentiation. If a given pathway is turned off during differentiation (e.g. the ability of progressing through mitosis), such a pathway will never be activated, whatever the level of stimulation. This, however, is only one of many possibilities. Other strategies may depend on the intensity or duration of the stimulation; i.e. given the same signalling molecule, a brief and intense signal can turn on a specific pathway, whereas a weak but prolonged stimulation leads to the activation of a different one (Dolmetsch et al 1998). The array of possibilities for diversifying the signal output includes many other alternatives such as interactions of multiple signalling pathways, temporal complexities (oscillations versus steady state increases etc; Li et al 1998). Over the last few years, another strategy has been the focus of much attention: the spatial organization of signalling molecules. In other words, not only can the quantitative and temporal aspects of signal generation ultimately regulate the output signal, but also the spatial organization within the cell can dictate the final response to the stimulus. For the fine mapping of the spatial complexity of second messengers however, new methodologies were required. The techniques developed are capable of monitoring — dynamically and with high spatial and temporal resolution — the concentration of these molecules within single living cells.

Spatial complexity of Ca^{2+} signalling

The measurement of cytoplasmic Ca^{2+} with fluorescent probes has been possible for several years (Zacharias et al 2000). The individual events leading to global Ca^{2+} elevations, however, have only relatively recently been revealed, through the use of high-speed confocal microscopy (Berridge et al 2000a, Berridge 2001 this volume). The changes in cytoplasmic Ca^{2+} are only part of the spatially complex Ca^{2+} signals. Within the lumen of various organelles (such as mitochondria, endoplasmic reticulum [ER], Golgi) changes also occur, and these are at least as important as those of the cytosol (Pozzan et al 1994). The main limitation of fluorescent dyes — either microinjected or loaded via membrane-permeable esters — is that these indicators reside in the cytoplasm and only in some cases (often erratically) do they end up within organelles. Information about the latter is thus largely indirect and obtained through the use of selective drugs. A more direct approach of monitoring the dynamics of organelle Ca^{2+} changes in living cells is that of selectively targeting a Ca^{2+} probe. This can be very efficiently and selectively done using molecularly engineered Ca^{2+}-sensitive proteins. At present, two families of such probes are available: those based on the Ca^{2+}-sensitive photoprotein aequorin (Robert et al 2000), and those based on mutants of green fluorescent protein (GFP; the so-called cameleons and camgaroos; Miyawaki et al 1999, Baird et al 1999).

Aequorin is a 20 kDa protein produced by the jellyfish *Aequorea victoria* that was used extensively in the 60s and 70s to probe Ca^{2+} in the cytoplasm of large cells, such as squid giant axons or muscle fibres (Blinks et al 1978). Functional aequorin contains a covalently linked coenzyme, named coelenterazine. Upon binding of Ca^{2+} to the specific sites (three EF hand sites/molecule) aequorin undergoes a conformational change that results in the emission of a photon and oxidation of the coenzyme. The major obstacle to the wide use of aequorin as a Ca^{2+} probe was the necessity of microinjecting it into the cytoplasm. Cloning of the cDNA encoding this protein (Inouye et al 1985) and modern molecular biology techniques have overcome this problem, and it is now possible to transfect any cell of interest; the expressed photoprotein is rendered functional simply by incubating the cells with exogenous coelenterazine. In addition, aequorin can be molecularly engineered with the insertion of targeting sequences that dictate the localisation of the expressed protein in specific cell compartments (reviewed in Robert et al 2000).

Whereas aequorin is naturally Ca^{2+}-sensitive, the GFP-based Ca^{2+} sensors have been molecularly tailored to this purpose. The so-called cameleons are chimeric proteins composed of four independent domains (Miyawaki et al 1999, 1997): two differently coloured GFPs are located at the N and C termini, while the central core is composed of calmodulin (CaM) and the CaM binding domain of myosin light chain kinase (M13); several variants of this basic structure have been generated by different groups (Romoser et al 1997, Persechini et al 1997). Ca^{2+}-sensing by cameleons depends on the phenomenon of fluorescence resonance energy transfer (FRET) between the two GFPs. Oversimplifying, FRET results from the non-radiant transfer of energy from an excited donor chromophore (e.g. BFP, a blue mutant of GFP) to an acceptor chromophore (e.g. GFP). This phenomenon is highly sensitive to the distance and orientation of the chromophores. In practical terms, the sample is excited at a wavelength typical of the donor (in the case mentioned above, UV light) and, if FRET occurs, the resulting emission is that typical of the acceptor (green light). When the $[Ca^{2+}]$ is low, FRET of cameleons is minimal, while binding of Ca^{2+} to CaM results in a drastic conformational change that increases FRET.

Camgaroos, on the other hand, result from the insertion of CaM at a specific site within the GFP molecule (Baird et al 1999). In this case, the conformational change induced by Ca^{2+} binding to the CaM domain changes the fluorescence intensity of the GFP. As in the case of aequorin, cameleons and camgaroos, being genetically encoded, can be transfected and targeted to specific cellular regions. At the moment these protein indicators can be targeted to the nucleus, mitochondria (matrix or intermembrane space), the ER, the Golgi, the surface of secretory vesicles or the inner leaflet of the plasma membrane; a chimeric aequorin for the lumen of secretory vesicles is also available (T. Pozzan & L. Filippin, unpublished data). To our knowledge, peroxisomes remain the only organelles for which a Ca^{2+} probe is still not available.

These probes have led to major improvements in our understanding of the spatial complexity of the Ca^{2+} signalling pathway. It is beyond the purpose of this brief report to extensively review this information and we will limit ourselves to a brief discussion of recent data regarding Ca^{2+} handling by mitochondria.

Mitochondria as biosensors of Ca^{2+} microheterogeneity

It has been known for a long time that mitochondria are endowed with the ability to accumulate Ca^{2+} at the expense of the membrane potential generated by the respiratory chain. This property, however, was generally attributed to experimental artefacts or even post-mortem effects — and there were reasons for this prejudice. For example, the apparent affinity of the mitochondrial Ca^{2+} uptake system is very low ($10–100\,\mu M$); i.e. well above the Ca^{2+} concentrations that are found in healthy living cells (for review see Pozzan & Rizzuto 2000a,b). Luckily, the unexpected finding that physiological stimuli result invariably in large increases in mitochondrial Ca^{2+} — more than 10-fold higher than the cytoplasmic values — renewed the interest in this phenomenon. The vast majority of the experts in the field currently agree with the idea that this apparent contradiction is explained by the microheterogeneity of Ca^{2+} within the cytoplasm. In particular, the idea has been put forward that mitochondria are strategically located close to the Ca^{2+} channels either of organelles (ER or sarcoplasmic reticulum [SR]) or of the plasma membrane (Pozzan & Rizzuto 2000a,b). At the mouth of these channels, the Ca^{2+} concentration is transiently much higher than in the rest of the cytoplasm; i.e. perfectly suited for the low-affinity Ca^{2+} transport into mitochondria. Thus, not only do mitochondria take up part of the Ca^{2+} released into the cytoplasm (exerting a significant buffering capacity on the amplitude of the cytoplasmic Ca^{2+} peaks), but they also play more subtle roles. These include, for example, modulating the feedback of Ca^{2+} on the channels, shaping the duration of the cytoplasmic Ca^{2+} increase, and local modulation of the activity of Ca^{2+}-dependent events, such as secretion (Montero et al 2000). One question that is still unanswered in this field is whether the proximity of mitochondria to the Ca^{2+} hot-spots is simply a stochastic event (mitochondria represent up to 20% of the cytoplasmic volume and, simply by random distribution, some of them are likely to be found in the vicinity of a Ca^{2+} channel), or whether some specific interaction is involved.

If the proximity of mitochondria to the Ca^{2+} release sites is a stochastic event — and considering that both the ER and mitochondria in a living cell move around continuously — one would expect that the individual mitochondria close to the Ca^{2+} release channels would continuously change. We have tried to address this issue by taking advantage of one of the characteristics of aequorin: upon binding of Ca^{2+}, aequorin emits a photon and is irreversibly 'consumed' (i.e. a second Ca^{2+}-binding event results in no photon emission). Given this premise, if we stimulate the cells twice with the same stimulus (one that induces the same increase in cytosolic Ca^{2+} concentration), different regions of the mitochondrial network would be exposed to the Ca^{2+} hot-spots during each challenge. In other words, the mitochondrial subpopulation exposed to a high Ca^{2+} microdomain would contain 'fresh' aequorin in both instances, and the measured average Ca^{2+} increase in mitochondria would therefore be identical. Conversely, if the regions of mitochondrial network that are close to the Ca^{2+}-release sites remain constant, the aequorin in that subpopulation of mitochondria is irreversibly and extensively consumed during the first stimulus — the apparent Ca^{2+} rise upon a second stimulation will, thus, be abolished or drastically reduced. Figure 1 shows that the latter alternative is correct. HeLa cells expressing mitochondrially targeted aequorin were challenged twice, with a 7 min interval, with histamine. The two stimuli resulted in identical cytoplasmic Ca^{2+} increases (not shown), while the first mitochondrial Ca^{2+} peak was almost threefold larger than the second. The data obtained in recent experiments support the concept that the subpopulation of the mitochondrial network that is close to the $InsP_3$-activated Ca^{2+} channels is

mtAEQ wt

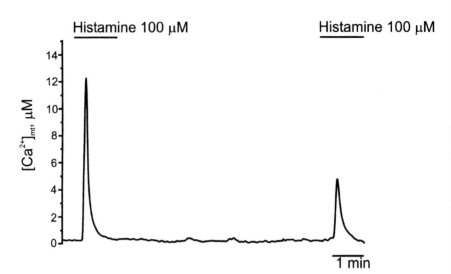

FIG. 1.   Effect on mitochondrial [Ca^{2+}] of two consecutive stimuli of histamine (100 μM) in HeLa cells transfected with aequorin targeted to mitochondria.

stably associated with those regions of the ER. The two histamine stimulations must be separated by over 60 min to obtain a complete recovery of the mitochondrial response. In other words, these experiments strongly suggest that the proximity of mitochondria to ER-released Ca^{2+} hot-spots is not a stochastic event, but it is due to a stable association between the two organelles. This finding obviously opens the search for the molecules responsible for this close association, but at the moment no candidates are available.

Cyclic AMP spatial heterogeneity

If the idea of spatial heterogeneity in the levels of Ca^{2+} is a widely accepted concept, much less is known about the other classical second messenger, cAMP. There are two reasons for the poor understanding of the spatial organization of cAMP signalling. The first is that the methodology for monitoring cAMP levels in single living cells has only recently become available. The second is that, unlike Ca^{2+}, the rate of diffusion of cAMP in living cells is thought to be very fast. Yet, indirect evidence pointing to the importance of localized increases of cAMP has been obtained, particularly in heart cells (Jurevicius & Fischmeister 1996). Two probes are currently available to monitor cAMP levels at the single-cell level and

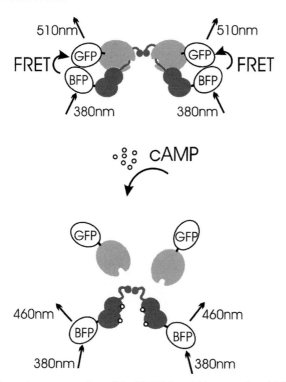

FIG. 2. Schematic representation of the FRET-based biosensor for cAMP. Incoming arrows indicate fluorescence excitation and outgoing arrows indicate fluorescence emission. Excitation and emission peak wavelengths are stated. The two subunits of protein kinase A are tagged with two variants of GFP suitable for FRET. The catalytic subunit of PKA is shown in light grey and the regulatory subunit of PKA is shown in dark grey. Donor and acceptor fluorophores are BFP and GFP, respectively. When cAMP is low, most PKA subunits form a GFP-tagged heterotetramer. In this condition, the donor and acceptor fluorophores are in close proximity and FRET occurs. When cAMP increases it binds to the regulatory subunits. The consequent conformational change in the regulatory subunit determines the release of the active catalytic subunit: the two fluorophores diffuse apart and FRET disappears.

they are both based on the same principle, FRET. The first of these probes was introduced nine years ago (Adams et al 1991) and is based on the covalent labelling of protein kinase A, PKA: the catalytic subunit (C) is labelled with fluorescein, whereas rhodamine is linked to the regulatory subunit (R). The second probe is genetically encoded and is based on the fusion of two differently coloured GFPs (suitable for FRET) to the PKA subunits (Zaccolo et al 2000). The rationale is the same in both cases: in conditions of low [cAMP], the C and R subunits of PKA are closely associated and FRET will be maximal. When [cAMP] rises, the C and R subunits dissociate and FRET is abolished. A scheme that summarizes this concept is presented in Fig. 2. The necessity to microinject covalently labelled PKA has limited enormously the use of the first probe, while the GFP-based probe can be introduced into living cells by transfection. This latter methodology has been introduced only a few months ago and accordingly only very few data are available. However, preliminary evidence obtained in our laboratory indicate that at least in some model systems—for example, cardiomyocytes in culture—it is possible to demonstrate spatially distinct domains of high cAMP upon stimulation of specific receptors. A crucial role in the maintenance of such steep local gradients of cAMP is played by the cAMP-degrading enzymes, as demonstrated by the homogeneous diffusion of cAMP throughout the cell upon pharmacological inhibition of phosphodiesterases.

Conclusions

In this brief report we mentioned a few methodological approaches to the study of the spatial complexity of two key second messengers — Ca^{2+} and cAMP — and we described some recent new data from our group. This is by no means an exhaustive review of the available methods to monitor the spatial complexity of these second messengers or of their downstream effector systems. The reader is referred to the reviews or to the original articles quoted below for more details. Our purpose was only that of stimulating the discussion on the novel aspects of signal transduction in the hope that some of the information provided might be of some help in modelling the spatial complexity of second messengers in living cells.

References

Adams SR, Harootunian AT, Buechler YJ, Taylor SS, Tsien RY 1991 Fluorescence ratio imaging of cyclic AMP in single cells. Nature 349:694–697

Baird GS, Zacharias DA, Tsien RY 1999 Circular permutation and receptor insertion within green fluorescent proteins. Proc Natl Acad Sci USA 96:11241–11246

Berridge MJ 2001 The versatility and complexity of calcium signalling. In: Complexity in biological information processing. Wiley, Chichester (Novartis Found Symp 239) p 52–67

Berridge MJ, Lipp P, Bootman MD 2000 Signal transduction. The calcium entry pas de deux. Science 287:1604–1605

Blinks JR, Mattingly PH, Jewell BR, van Leeuwen M, Harrer GC, Allen DG 1978 Practical aspects of the use of aequorin as a calcium indicator: assay, preparation, microinjection, and interpretation of signals. In: DeLuca MA (ed) Methods in enzymology, vol 57: Bioluminescence and chemiluminescence. Academic Press, New York, p 292–328

Dolmetsch RE, Xu K, Lewis RS 1998 Calcium oscillations increase the efficiency and specificity of gene expression. Nature 392:933–936

Inouye S, Noguchi M, Sakaki Y et al 1985 Cloning and sequence analysis of cDNA for the luminescent protein aequorin. Proc Natl Acad Sci USA 82:3154–3158

Jurevicius J, Fischmeister R 1996 cAMP compartmentation is responsible for a local activation of cardiac Ca^{2+} channels by beta-adrenergic agonists. Proc Natl Acad Sci USA 93:295–299

Li W, Llopis J, Whitney M, Zlokarnik G, Tsien RY 1998 Cell-permeant caged InsP3 ester shows that Ca^{2+} spike frequency can optimize gene expression. Nature 392:936–941

Miyawaki A, Llopis J, Heim R et al 1997 Fluorescent indicators for Ca^{2+} based on green fluorescent proteins and calmodulin. Nature 388:882–887

Miyawaki A, Griesbeck O, Heim R, Tsien RY 1999 Dynamic and quantitative Ca^{2+} measurements using improved cameleons. Proc Natl Acad Sci USA 96:2135–2140

Montero M, Alonso MT, Carnicero E et al 2000 Chromaffin-cell stimulation triggers fast millimolar mitochondrial Ca^{2+} transients that modulate secretion. Nat Cell Biol 2:57–61

Persechini A, Lynch JA, Romoser VA 1997 Novel fluorescent indicator proteins for monitoring free intracellular Ca^{2+}. Cell Calcium 22:209–216

Pozzan T, Rizzuto R 2000a High tide of calcium in mitochondria. Nat Cell Biol 2:E25–E27

Pozzan T, Rizzuto R 2000b The Renaissance of Mitochondrial Calcium Transport. Eur J Biochem 267:5269–5273

Pozzan T, Rizzuto R, Volpe P, Meldolesi J 1994 Molecular and cellular physiology of intracellular calcium stores. Physiol Rev 74:595–636

Robert V, Pinton P, Tosello V, Rizzuto R, Pozzan T 2000 Recombinant aequorin as a tool for monitoring calcium concentration in subcellular compartments. In: Thorner J, Emr SD, Abelson JN, Simon MI (eds) Methods in enzymology, vol 327: Applications of chimeric genes and hybrid proteins, part B: cell biology and physiology. Academic Press, San Diego, p 440–456

Romoser VA, Hinkle PM, Persechini A 1997 Detection in living cells of Ca^{2+}-dependent changes in the fluorescence emission of an indicator composed of two green fluorescent protein variants linked by a calmodulin-binding sequence. A new class of fluorescent indicators. J Biol Chem 272:13270–13274

Zaccolo M, De Giorgi F, Cho CY et al 2000 A genetically encoded, fluorescent indicator for cyclic AMP in living cells. Nat Cell Biol 2:25–29

Zacharias DA, Baird GS, Tsien RY 2000 Recent advances in technology for measuring and manipulating cell signals. Curr Opin Neurobiol 10:416–421

DISCUSSION

Kahn: Yesterday we talked a bit about the difference between graded responses and quantal responses. Using the imaging approach in both the cAMP and the Ca^{2+} system, at the level of single cells if you see 50% stimulation, is this because half the mitochondria show a response and half don't? Or does each of the mitochondria show a 50% response?

Pozzan: My interpretation is that in each single cell, 50% of the mitochondria respond, but I cannot prove this hypothesis formally. In other words, according to my interpretation, 50% of the mitochondria of each cell show a big spike and 50% hardly respond. This interpretation is consistent with the very slow recovery of the mitochondrial response, despite the fact that the Ca^{2+} rise in the cytoplasm is unaffected.

Brabant: Is this a functional regulation of the non- or less-responding mitochondria, or is it a dynamic state fluctuating from high to low?

Pozzan: This is a difficult question to answer. The possibility that mitochondria are close to the Ca^{2+} channels simply as a consequence of a stochastic event is, in my opinion, too risky for the cell, because this process is vital in generating ATP. There is evidence in some specific cell types that Ca^{2+} release always starts from the same sites. There are InsP$_3$ receptor clusters that are more sensitive than others, but, at the moment, it is not known whether or not mitochondria cluster preferentially around these receptors.

Sejnowski: In this regard, is the Ca^{2+} transport localized?

Pozzan: Nothing is known at the molecular level about the components of the mitochondrial Ca^{2+} uptake or release mechanisms.

Laughlin: Does this mean that we know very little about how the mitochondria return the Ca^{2+}?

Pozzan: To a large extent under physiological conditions they return Ca^{2+} to the cytoplasm through the action of two exchangers, the Na$^+$/Ca^{2+} and the H$^+$/Ca^{2+} electroneutral exchangers. Under physiological conditions these are the only two Ca^{2+} efflux pathways that mediate Ca^{2+} release from mitochondria. This conclusion is based on the use of drugs that are reasonably specific for these exchangers. They either reduce the speed or block Ca^{2+} efflux. Under other conditions there is a third pathway, the so called 'permeability transition pore'. If this pore opens, Ca^{2+} can come out through it very rapidly. The conditions under which this pore opens are usually very artificial and the evidence that it does so in healthy cells is very limited.

Laughlin: Presumably the rate at which Ca^{2+} is returned has an influence on the dynamics and the efficiency of signalling.

Pozzan: In terms of cell energy balance, we know that when Ca^{2+} is taken up, the mitochondria stop making ATP. However, the Ca^{2+} microdomain mechanism is economically advantageous for the cells. Mitochondria stop making ATP for just a few milliseconds, the time required for the rapid uptake of Ca^{2+}. The Ca^{2+} microdomains disappear rapidly and the efflux pathway is very slow. In other words, by sacrificing a few hundred milliseconds of ATP synthesis, this arrangement ensures that the mitochondrial dehydrogenases remain activated for several minutes (and thus ATP synthesis is increased for a long period).

Schultz: Why is it so difficult to set up a single cell assay for cGMP similar to the one for cAMP? Is it possible in the cAMP-dependent protein kinase to replace the cAMP binding domain with a cGMP binding domain?

Pozzan: It could be done. We haven't tried. The rumours are that a few groups have tried to do the same with cGMP dependent kinase. I haven't seen anything published on this, which probably means that it hasn't worked. We considered the alternative possibility, given that this construct works, i.e. to substitute the cAMP binding domain with the cGMP binding domain, hoping that the rest of the structure would remain unperturbed.

Brenner: This will work: I have made the construct, although not for this purpose. Just replace it. It is one PCR experiment.

Sejnowski: But you haven't published it.

Brenner: No. This was done in bacteria, for another purpose.

Berridge: I was struck by the very slow recovery when you took out 10% of the sites. This didn't fit too well with the idea that these are sausage-shaped structures that are in contact with 90% of the normal aequorin, which should diffuse in quite quickly. It seems rather a slow process for this kind of tunnelling.

Pozzan: The 10% I was talking about are regions where the vicinity of the mitochondria to the ER could not be resolved. In these regions the two organelles are less than 80 nm apart. My idea is that Ca^{2+} has privileged entry sites in those areas, but then it diffuses through the tubes intralumenally. Thus the amount of aequorin that will be burned will be much larger than 10%, depending on how far Ca^{2+} diffuses into the mitochondrial tubular network.

Berridge: This sieve idea is interesting, because it fits nicely with the possibility that you have a mitochondrion sitting close to the channel, which takes the Ca^{2+} away and prevents the negative feedback on the channel, and then it diffuses down through the mitochondria and gets sprayed into the cytoplasm at deeper points.

Schöfl: Do you have any indications from your single cell measurements of cAMP concentrations whether cAMP rises in a graded or amplitude-regulated fashion? Are there any temporal fluctuations in cAMP concentrations like in Ca^{2+} signalling systems?

Pozzan: These are all planned experiments. If I had to bet, I would bet that cAMP will oscillate as well, but with less sharp fluctuations than those of Ca^{2+}. It appears unlikely to me that cAMP will not oscillate in some kind of synchrony with Ca^{2+}, given that both the cyclases and the phosphodiesterases are Ca^{2+} modulated.

Humoral coding and decoding

Klaus Prank*, Martin Kropp† and Georg Brabant†

*Research and Development, BIOBASE Biological Databases/Biologische Datenbanken GmbH, Mascheroder Weg 1b, D-38124 Braunschweig, Germany and †Computational Endocrinology Group, Department of Clinical Endocrinology, Medical School Hanover, Carl-Neuberg-Str. 1,D-30625 Hanover, Germany

Abstract. Humoral communication systems are dynamically regulated. Most hormones are released in a pulsatile or burst-like manner into the bloodstream. It is well known that information coded in the frequency and amplitude of secretory pulses allows for the differential regulation of specific target cell function and structure. However, despite intensive study of transmembrane signalling relatively little is known about how the temporal dynamics of extracellular humoral stimuli specifically regulates the temporal pattern of intracellular signalling pathways, such as Ca^{2+}-dependent signalling. Repetitive spikes of Ca^{2+} encode this information in their amplitude, duration and frequency, and are in turn decoded into the pattern of gene expression and phosphorylation of target proteins. Using a mathematical model for G protein-coupled Ca^{2+} signalling and information-theoretic approaches to stimulus reconstruction we have systematically quantified the amount of information coded in the Ca^{2+}-signal about the dynamics of the stimulus, which allows us to explore the temporal bandwidth of transmembrane signalling. These *in silico* approaches permit us to differentiate the amount of information coded in the frequency, temporal precision, amplitude and the complete Ca^{2+}-signal. This may open an avenue to the quantification of information flow and processing in the intra- and intercellular coding and decoding machinery.

2001 Complexity in biological information processing. Wiley, Chichester (Novartis Foundation Symposium 239) p 96–110

Biological information transfer

The two major biological systems that communicate information over long distances are the nervous and endocrine systems. There are common principles of information processing in humoral and neuronal communication. Both systems are regulated dynamically. Information transferred via the nervous system appears to be predominantly encoded in the frequency of action potentials (frequency coding, Adrian 1928). The same holds true for humoral signalling, which is conventionally viewed as an analogue operating system. However, most hormones are secreted in a burst-like or pulsatile manner into the bloodstream and modulations in the amplitude- and/or frequency of secretory pulses are able to specifically regulate the function and structure of distinct target organs (Brabant et al 1992). A number of examples support the theory that disruptions of the temporal pattern of secretion serve as the basis of endocrine diseases (Brabant et al 1992). Compared to the millisecond time scale of neuronal information processing, the time scale of hormonal rhythms is at least five orders of magnitude slower. The period of these hormonal rhythms ranges from minutes (for hormones such as insulin and catecholamines regulating acute physiological processes), to hours (circadian and infradian rhythms) (Brabant et al 1992).

Transmembrane signal transduction in humoral signalling

The frequency coding scheme seen in the pulsatile pattern of hormone secretion continues across the cell membrane through G protein-coupled receptor signalling to the temporal pattern of intracellular Ca^{2+} dynamics (Woods et al 1986). The ubiquitous intracellular second messenger Ca^{2+} has been demonstrated to be organized in complex spatiotemporal patterns. These patterns exhibit repetitive spikes or oscillations as well as waves in a variety of different cell types upon stimulation with hormonal agonists and neurotransmitters (Berridge 1993). The frequency of Ca^{2+} oscillations is modulated by the dose of humoral agonists (Woods et al 1986) in analogy to the modulation of the frequency of neuronal firing by the depolarization current (Adrian 1928). In addition, it has been demonstrated experimentally that square wave stimuli of α_1-adrenergic agonists mimicking the physiological pattern of pulsatile catecholamine secretion lead to a modulation of the Ca^{2+} spike amplitude in single hepatocytes (Schöfl et al 1993). The modulation of the frequency, amplitude, and duration of Ca^{2+} spikes plays an important role in the regulation of intracellular processes and can be decoded in the activation of enzymes such as the ubiquitous Ca^{2+}–calmodulin-dependent protein kinase II (De Koninck & Schulman 1998) or the activation of gene expression through the regulation of transcription factors (Dolmetsch et al 1997).

Temporal coding

For neuronal information processing it has been demonstrated that the exact timing of individual spikes (temporal coding) is relevant to characterize the neuronal response (Rieke et al 1997). This contribution to the information content of a neuronal spike train is neglected if only the firing rate (frequency coding) is taken into account. The idea behind temporal coding is that the timing of spikes plays an important role in encoding various aspects of the stimulus as has been demonstrated in a number of different sensory neuronal systems (Chung et al 1970, Abeles 1990, Strehler & Lestienne 1986, Bialek et al 1991, Eskandar et al 1992, Singer & Gray 1995, Decharms & Merzenich 1996, Laurent 1996, Wehr & Laurent 1996, Gabbiani et al 1994, Haag & Borst 1997, Lisman 1997). Temporal coding allows for a significant increase in the information capacity of neuronal spike trains. One well-studied example for the use of temporal coding is the visual information processing in the fly (Bialek et al 1991). Flies control their flight behaviour visually on a time scale of tens of milliseconds, which allows them to observe only a few action potentials. Thus, their behavioural decisions cannot be based on frequency coding but rather on temporal coding. Experimentally, the neuronal spike train of a movement-sensitive neuron (H1 neuron) was recorded upon visual stimuli consisting of the angular velocity of a moving random pattern. Using an information-theoretic approach for stimulus reconstruction it is possible to reconstruct the dynamics of stimulus from the neuronal spike train. This allows for an estimation of the coding efficiency, information rate, and temporal bandwidth of visual information processing (Bialek et al 1991).

Simulation of transmembrane signal transduction

In analogy to neuronal signalling we addressed the question how temporal coding contributes to the encoding of humoral stimulus dynamics in the Ca^{2+} spike train. We used a mathematical model for the transduction of extracelluar fluctuating hormonal stimuli into Ca^{2+} spike trains (Chay et al 1995) which is based on experimental Ca^{2+} data in hepatocytes stimulated with an α_1-adrenergic pulsatile stimulus (Schöfl et al 1993). This model allows for the simulation of receptor-controlled activation of G proteins upon agonist stimulation, the subsequent activation of phospholipase C (PLC), and the build up of inositol-(1,4,5)-trisphosphate ($InsP_3$) which finally triggers the release of Ca^{2+} from internal stores such as the endoplasmic reticulum (ER). The level of cytosolic Ca^{2+} drops fast as Ca^{2+} is pumped back into the ER leading to repetitive spikes of Ca^{2+} (Fig. 1). To explore the temporal bandwidth of transmembrane signal transduction we generated fluctuating stimuli from Gaussian white noise by low-pass filtering as a first approximation of the physiological input of pulsatile α_1-adrenergic stimulation (Fig. 2).

Stimulus reconstruction from Ca^{2+} spike trains

The use of a reverse-engineering approach allows us to compute a temporal filter that, when convolved with a spike train in response to the stimulus, will produce an estimate of the stimulus. By using such an information-theoretic approach part of the temporal dynamics of the stimulus can be reconstructed from the Ca^{2+} spike train and the rate and accuracy of information transmission can be estimated (Bialek et al 1991, Rieke et al 1993, Gabbiani & Koch 1996, Gabbiani 1996, Prank et al 1998b, 2000). A linear estimate of the stimulus is calculated by

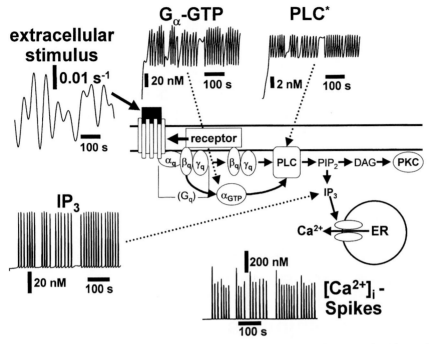

FIG. 1.   Simulation of transmembrane signal transduction. The simulations are based a model of receptor-controlled Ca^{2+} oscillations (Chay et al 1995). DAG, diacylglycerol; IP_3, inositol-1,4,5-trisphosphate; PIP_2, phosphatidylinositol-4,5-bisphosphate; PLC, phospholipase C.

convolving the spike train with a filter. The filter is chosen in such a way as to minimize the mean square error between the stimulus and the estimate. The filter is not causal in the sense that the occurrence of a spike can be used to predict the future temporal dynamics of the stimulus. This is of course only possible because of the response properties of the simulated cell. Once the best linear estimate is found, the 'noise' contaminating the reconstruction is defined as the difference between the estimated stimulus and the stimulus (Fig. 2). There are two measures to quantify the accuracy of the reconstruction and thus the information transmitted from the stimulus respectively. The coding fraction represents the percentage of temporal stimulus fluctuations encoded in units of the standard deviation of the stimulus. The coding fraction takes a maximum value of 1 when the stimulus is perfectly estimated and the minimum value of 0 if the stimulus estimation from the Ca^{2+} spike train is at chance level (Gabbiani 1996, Gabbiani & Koch 1996). An alternative measure, the mutual information transmitted by the reconstructions about the stimulus can be used (Bialek et al 1991). For a Gaussian white noise stimulus this is a measure of the equivalent rate of information

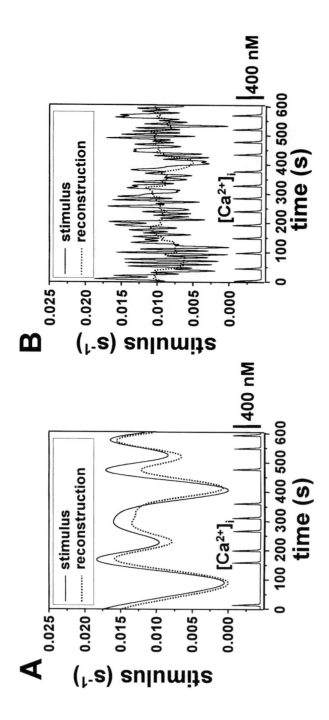

FIG. 2. Stimulus reconstruction. Upper traces: original and reconstructed stimulus; lower trace: corresponding Ca^{2+} spike train. (A) Regular stimulus with low frequency content. Cut-off frequency of the low-pass filter $f_c = 10$ mHz. (B) Irregular stimulus with high frequency content ($f_c = 100$ mHz).

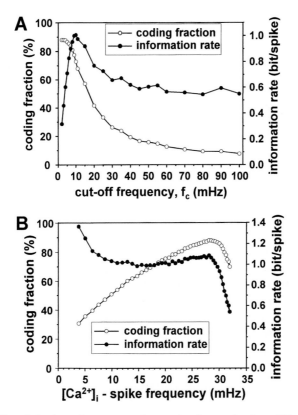

FIG. 3. Coding behaviour in transmembrane signal transduction. (A) Impact of the bandwidth of the stimulus on coding fraction and information rate. (B) Impact of the mean Ca^{2+} spike frequency on coding fraction and information rate.

transmission. A lower bound for the rate of information transmitted per Ca^{2+} spike is obtained by dividing the rate of information transmission by the mean Ca^{2+} spike frequency. More details on the algorithms and software used for the reconstruction can be found in Prank et al (1998b, 2000) and Gabbiani & Koch (1998).

Effect of the stimulus bandwidth on the coding behaviour

We used stimuli with low frequency as well as high frequency content by choosing filters with different cut-off frequencies. Figure 3A demonstrates the close agreement between the original and estimated stimulus for the low cut-off frequency resulting in a high values for the coding fraction as well as information rate and a poor reconstruction of the stimulus with the high frequency content stimulus leading to a smoothed moving average estimate (Fig. 3B). Thus, the bandwidth of this stimulus seems to be beyond the bandwidth of the transmembrane coding machinery. This becomes obvious in the temporal pattern of the corresponding Ca^{2+} spike train. The spacing of Ca^{2+} spikes for the irregular stimulus is very uniform and does not allow for coding the stimulus dynamics in the temporal pattern. In contrast, the regular stimulus with low frequency content results in a Ca^{2+} spike train exhibiting a large variability in the interspike intervals which enables coding of the stimulus dynamics in the timing of Ca^{2+} spikes. This has been systematically evaluated by changing the cut-off frequency from 3 mHz to 100 mHz producing low-to-high frequency content stimuli (Fig. 3A). Increasing the cut-off frequency led to a monotonic decrease of the coding fraction, whereas the information transmitted per spike remained constant for cut-off frequencies larger than approximately 30 mHz. Since the mean Ca^{2+} spike frequency has an effect on the coding behaviour, we chose a fixed cut-off frequency of 10 mHz to generate a stimulus which has been demonstrated to yield good reconstructions of the stimulus. The $[Ca^{2+}]_i$-spike frequency was increased monotonically by increasing the amplitude range of the fluctuating stimuli (Fig. 3B). The maximum of the coding fraction and the information transmitted per Ca^{2+} spike were 0.87 and 1.1 bit/spike respectively at a mean Ca^{2+} spike frequency of 27 mHz. At low Ca^{2+} spike frequencies below 10 mHz the coding fraction decreased below 0.5, in contrast to the information transmitted per spike which increased to 1.4 bit/spike (Fig. 3B).

Coding in Ca^{2+} spike amplitude and interspike interval

Since Ca^{2+} spikes can be modulated not only in their frequency and temporal pattern but also in their amplitude, we investigated how the amplitude and interspike interval (ISI) might differentially code information about the stimulus dynamics. To address this question we used the Ca^{2+} spike signal containing ISI and amplitude information and compared this to the Ca^{2+} signal containing only ISI information by 'clamping' the amplitude to a fixed value (Fig. 4). The reduction in the coding performance by 'clamping' the Ca^{2+} spike amplitude is exemplified in Fig. 4. The coding fraction reduces from 73% to 62% and the information transmitted per spike from 1.0 bit to 0.7 bit.

Universality in biological information transfer

Both long-range signalling systems — neuronal and endocrine systems — are operating on time scales that differ by three-to-five orders of magnitude, as indicated on the logarithmic scale of the y-axis for the information rate in bits/spike (Fig. 5). However, if we determine the information rate per action potential or per Ca^{2+} spike, they are of the same order, ranging from 1–4 bit/spike in neuronal signalling and 1 bit/spike for the model system for humoral signalling

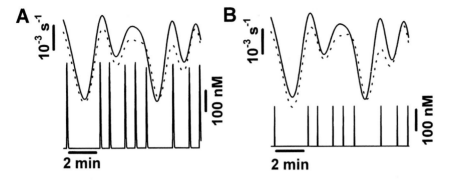

FIG. 4. Coding in Ca^{2+}-spike amplitude and interspike interval (ISI). (A) Amplitude and ISI information, coding fraction = 73%, information rate = 1.0 bit/spike. (B) only ISI information, coding fraction = 62%, information rate = 0.7 bit/spike.

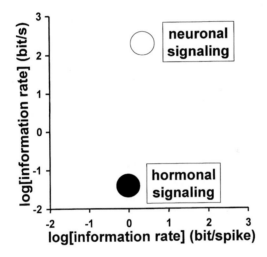

FIG. 5. Universality in biological information transfer.

investigated in this study. This suggests some sort of universality principle in information processing between neuronal and humoral signalling.

Discussion

It has been demonstrated in recent years that studying the average firing rate in neuronal signalling neglects most of the information contained in a neuronal spike train. Recently, the temporal coding of information in the patterns of spikes, both in the single cell as well as between multiple cells, has received renewed attention. The broad idea that spike timing plays an important role in encoding various aspects of the stimulus, in particular across an ensemble of cells, is supported by experiments in a variety of sensory systems such as locust olfaction, electric fish electrosensation, cat vision and olfaction, and monkey vision and audition (Chung et al 1970, Abeles 1990, Strehler & Lestienne 1986, Bialek et al 1991, Eskandar et al 1992, Singer & Gray 1995, Decharms & Merzenich 1996, Laurent 1996, Wehr & Laurent 1996, Gabbiani et al 1994, Lisman 1997). Because little or no information can be encoded into a stream of completely regularly spaced action potentials, this raises the question of how variable neuronal firing really is and what the relation is between variability and the neural code (Rieke et al 1997, Mainen & Sejnowski 1995). It is the mathematical theory of stochastic point processes and the field of statistical signal processing that offer us the adequate tools for attacking these questions.

For humoral information processing, the coding of information in the extracellular pattern of pulsatile hormone secretion and the temporal dynamics of intracellular second messengers such as Ca^{2+} have been studied mainly regarding their mean frequency and amplitude. It is not known yet whether temporal coding (i.e. the timing of single secretory pulses as well as Ca^{2+} spikes besides the well known frequency) and/or amplitude coding schemes play a physiological role in the regulation of intracellular signalling and target cell function and structure, respectively. The amplitude (AM) and frequency modulation (FM) of Ca^{2+} spike trains have been reported to regulate distinct cellular processes differentially (Berridge 1997). The FM mode of Ca^{2+} signalling is used to control processes such as secretion (Rapp & Berridge 1981), glycogen metabolism in hepatocytes (Woods et al 1986, Schöfl et al 1993), and differentiation in the neuronal system (Gu & Spitzer 1995, Gomez & Spitzer 1999). The AM mode and duration of the Ca^{2+} signal on the other hand allow for differential gene activation in B lymphocytes (Dolmetsch et al 1997). An additional example for the FM regulation of cellular signalling by calcium is the ubiquitous Ca^{2+}–calmodulin-dependent protein kinase II (CaM kinase II). *In silico* approaches have proposed that CaM kinase acts as a molecular frequency decoder of Ca^{2+} spikes (Hanson et al 1994, Dosemeci & Albers 1996, Prank et al 1998a). These results are confirmed in *in vitro* experiments (De Koninck & Schulman 1998). This versatility and complexity of Ca^{2+} signalling from elementary events to global signals is elaborated in Berridge (2001, this volume), demonstrating the wide range of cellular mechanisms controlled by this ubiquitous second messenger.

In addition to the versatility and complexity already known for calcium signalling, we have used an *in silico* approach to explore temporal coding in the encoding and decoding of dynamic humoral stimuli through Ca^{2+} signalling. The information content of Ca^{2+} spike trains about the extracellular stimulus dynamics was quantified by two different measures: the coding fraction and the rate of information transmitted per Ca^{2+} spike. Up to 90% of the stimulus dynamics could be coded in the temporal pattern of a Ca^{2+} spike train with a maximum information rate of 1.1 bit/spike. Although the coding fraction severely dropped with a decrease in the mean Ca^{2+} spike frequency or an increase of the cut-off frequency of the stimulus, the information rate per Ca^{2+} spike kept relatively constant over a broad range of cut-off frequencies and Ca^{2+} spike frequencies. Increasing the frequency content of the stimulus to yield highly irregular stimuli led to almost regular Ca^{2+} spike trains with poor reconstructions of the original stimulus. These results clearly demonstrate that temporal coding is capable of increasing the information capacity compared to codes that only rely on the mean firing rate. Besides temporal coding in the pattern of action potentials and Ca^{2+} spikes, Ca^{2+} signalling allows for a differential coding of information in the amplitude and interspike intervals. This has been demonstrated by artificially 'clamping' the Ca^{2+} spike amplitude to a fixed value leading to a reduction in the coding fraction as well as the information transmitted per Ca^{2+} spike.

However, it would be a challenge to test these results experimentally by using time-varying hormonal stimuli matching the physiological pattern of pulsatile hormonal secretion to explore the encoding and decoding machinery of transmembrane signalling. Although we have demonstrated that cells are capable of increasing their information processing capacity by making use of temporal as well as amplitude coding, it remains an open issue to determine whether they are making really use of it and what the biological meaning of the temporal code would be. The information-theoretic approaches introduced to study neuronal information processing (Bialek et al 1991) which we adapted for calcium signalling (Prank et al 1998b, 2000) might be used in the future to quantify the dynamics of the information flow and the operational time scales in other parts of cellular signalling, i.e. between cells as well as between parallel pathways ('cross-talk'). The information-theoretic measures might then be correlated with cellular responses such as the level of phosphorylation of proteins or levels of gene expression to give the different codes a biological meaning.

Acknowledgement

This work was supported by the Deutsche Forschungsgemeinschaft under grant Br 915/4-4 and Bra 915/9-1.

References

Abeles M 1990 Corticonics: neural circuits of the cerebral cortex. Cambridge University Press, Cambridge

Adrian ED 1928 The basis of sensation: the action of the sense organs. W.W. Norton, New York

Berridge MJ 1993 Inositol trisphosphate and calcium signalling. Nature 361:315–325

Berridge MJ 1997 The AM and FM of calcium signalling. Nature 386:759–760

Berridge MJ 2001 The versatility and complexity of calcium signalling. In: Complexity in biological information processing. Wiley, Chichester (Novartis Found Symp 239) p 52–67

Bialek W, Rieke F, de Ruyter van Steveninck RR, Warland D 1991 Reading a neural code. Science 252: 1854–1857

Brabant G, Prank K, Schöfl C 1992 Pulsatile patterns in hormone secretion. Trends Endocrinol Metab 3:183–190

Chay TR, Lee YS, Fan YS 1995 Appearance of phase-locked Wenckebach-like rhythms, devil's staircase and universality in intracellular calcium spikes in non-excitable cell models. J Theor Biol 174:21–44

Chung SH, Raymond SA, Lettvin JY 1970 Multiple meaning in single visual units. Brain Behav Evol 3:72–101

De Koninck P, Schulman H 1998 Sensitivity of CaM kinase II to the frequency of Ca^{2+} oscillations. Science 279:227–230

deCharms RC, Merzenich MM 1996 Primary cortical representation of sounds by the coordination of action potential timing. Nature 381:610–613

Dolmetsch RE, Lewis RS, Goodnow CC, Healy JI 1997 Differential activation of transcription factors induced by Ca^{2+} response amplitude and duration. Nature 386:855–858 (erratum: 1997 Nature 388:308)

Dosemeci A, Albers RW 1996 A mechanism for synaptic frequency detection through autophosphorylation of CaM kinase II. Biophys J 70:2493–2501

Eskandar EN, Richmond BJ, Optican LM 1992 Role of inferior temporal neurons in visual memory. I. Temporal encoding of information about visual images, recalled images, and behavioral context. J Neurophysiol 68:1277–1295

Gabbiani F 1996 Coding of time-varying signals in spike trains of linear and half-wave rectifying neurons. Network: Comput Neur Syst 7:61–85

Gabbiani F, Koch C 1996 Coding of time-varying signals in spike trains of integrate-and-fire neurons with random threshold. Neur Comput 8:44–66

Gabbiani F, Koch C 1998 Principles of spike train analysis. In: Koch C, Segev I (eds) Methods in neuronal modelling, 2nd edn: from ions to networks. MIT Press, Cambridge, MA, p 313–360

Gabbiani F, Midtgaard J, Knöpfel T 1996 Synaptic integration in a model of cerebellar granule cells. J Neurophysiol 72:999–1009 (erratum: 1996 J Neurophysiol 75)

Gomez TM, Spitzer NC 1999 In vivo regulation of axon extension and pathfinding by growth-cone calcium transients. Nature 397:350–355

Gu X, Spitzer NC 1995 Distinct aspects of neuronal differentiation encoded by frequency of spontaneous Ca^{2+} transients. Nature 375:784–787

Haag J, Borst A 1997 Encoding of visual motion information and reliability in spiking and graded potential neurons. J Neurosci 17:4809–4819

Hanson PI, Meyer T, Stryer L, Schulman H 1994 Dual role of calmodulin in autophosphorylation of multifunctional CaM kinase may underlie decoding of calcium signals. Neuron 12:943–956

Laurent G 1996 Dynamical representation of odors by oscillating and evolving neural assemblies. Trends Neurosci 19:489–496

Lisman JE 1997 Bursts as a unit of neural information: making unreliable synapses reliable. Trends Neurosci 20:38–43

Mainen ZF, Sejnowski TJ 1995 Reliability of spike timing in neocortical neurons. Science 268:1503–1506

Prank K, Läer L, von zur Mühlen A, Brabant G, Schöfl C 1998a Decoding intracellular calcium spike trains. Europhys Lett 42:143–147

Prank K, Schöfl C, Läer L, Wagner M, von zur Mühlen A, Brabant G, Gabbiani F 1998b Coding of time-varying hormonal signals in intracellular calcium spike trains. In: Altman RB, Dunker

AK, Hunter L, Klein TE (eds) Biocomputing '98, Proceedings of the Pacific Symposium. World Scientific, Singapore, p 633–644

Prank K, Gabbiani F, Brabant G 2000 Coding efficiency and information rates in transmembrane signaling. Biosystems 55:15–22

Rapp PE, Berridge MJ 1981 The control of trasnepithelial potential oscillations in the salivary gland of calliphora erythrocephala. J Exp Biol 93:119–132

Rieke F, Warland D, Bialek W 1993 Coding efficiency and information rates in sensory neurons. Europhys Lett 22:151–156

Rieke F, Warland D, de Ruyter van Steveninck R, Bialek W 1997 Spikes: exploring the neural code. MIT Press, Cambridge, MA

Schöfl C, Brabant G, Hesch RD, von zur Mühlen A, Cobbold PH, Cuthbertson KS 1993 Temporal patterns of alpha 1-receptor stimulation regulate amplitude and frequency of calcium transients. Am J Physiol 265:C1030–C1036

Singer W, Gray CM 1995 Visual feature integration and the temporal correlation hypothesis. Annu Rev Neurosci 18:555–586

Strehler BL, Lestienne R 1986 Evidence on precise time-coded symbols and memory of patterns in monkey cortical neuronal spike trains. Proc Natl Acad Sci USA 83:9812–9816

Wehr M, Laurent G 1996 Odour encoding by temporal sequences of firing in oscillating neural assemblies. Nature 384:162–166

Woods NM, Cuthbertson KSR, Cobbold PH 1986 Repetitive transient rises in cytoplasmic free calcium in hormone-stimulated hepatocytes. Nature 319:600–602

DISCUSSION

Aertsen: This method of reconstruction very much focuses on the linear aspects of the system you are studying. The stronger the non-linearities, the lower the values of the coding fraction are bound to be. Does the fact that you get reasonably high numbers mean that the biological system is indeed linear? Or does it mean that your model is effectively linear? If so, is your model adequate to describe the biological system?

Prank: The stimulus dynamics could be reconstructed nearly optimally by our approach using a linear filter leading to values for the coding fraction of up to 90%. Thus, non-linear filters might not improve the stimulus reconstruction substantially. However, the mathematical model simulating transmembrane signal transduction in this study is based on coupled non-linear differential equations.

Laughlin: How well do the dynamics of the random hormone signal that you used in the model correspond to natural hormone signals?

Prank: We simulate the signal transduction on the basis of experimental data for an α_1-adrenergic stimulus. The dynamics, for example of catecholamines, correspond well at least with the low frequency, regular situation.

Berridge: I have a feeling that that is an erroneous depiction of what is actually happening in a liver cell. If you used stepwise concentration of agonist, which might be closer to reality, you would see no change in amplitude. This was the original experiment done by Peter Cobbold, which clearly showed that while frequency varied, amplitude remained constant.

Prank: It is a matter of debate: what does the physiological pattern of the hormonal stimulus look like? We know for most hormones that they are released in a burst-like or pulsatile manner (Brabant et al 1992). This holds true on a short time-scale as in the release of insulin with a mean interval between secretory pulses of around 12 min, and also for catecholamines on an even faster time scale of only a few minutes. With regard to the relevance of the fluctuating stimuli used in our study and the square wave pulses used in the experiments of Schöfl et al (1993), they are a first approximation of the physiological pattern of catecholamine release.

Berridge: What would be much more interesting would be to present the stimulus as a ramp.

Prank: That depends on the timescale of the increase. In our situation we are dealing with very fast increases.

Berridge: They are not fast relative to the signalling system inside the cell. The slowest rate of spiking in a liver cell is once every minute, and the highest rate is about once every second. The administration of the hormone would have to be within this time frame, and I don't believe that this is a very physiological simulation.

Schöfl: In the experiments shown by Klaus Prank we used the α_1-receptor-activated intracellular Ca^{2+} signal in hepatocytes to test whether different patterns of a pulsatile or burst-like activation of the receptor would result in distinct changes in the intracellular Ca^{2+} signal (Schöfl et al 1993). As the α_1-receptor is physiologically activated by noradrenaline, which is released from nerve endings in the liver, rapid and short bursts of noradrenaline rather than relatively slow changes in the agonist concentration (over several minutes) could be assumed. We therefore designed a perfusion system, which allowed for rapid changes in the agonist concentration at the site of the cell with a time constant of about 4 s to reach complete equilibration of the superfusate. With this system we were then able to demonstrate, that changes in the temporal pattern of α_1-receptor activation could lead to marked changes in the amplitude of the intracellular Ca^{2+} transients (Schöfl et al 1993). This is a rather unexpected phenomenon, since constant cell stimulation with different agonist concentrations only caused changes in the frequency of the Ca^{2+} transients with a more or less constant amplitude (Woods et al 1986). Our results were therefore compatible with the hypothesis that intercellular information could be encoded in the temporal pattern of neurotransmitter or hormonal stimulation of target cells. As pulsatile or burst-like secretion of hormones or neurotransmitters is the rule rather than the exception, this might be physiologically relevant. Furthermore, in subsequent experiments it could be shown that cross-signalling with the cAMP-signalling cascade can also cause marked changes in the amplitude as well as in the frequency of cytosolic Ca^{2+} transients. Interestingly, these changes in the intracellular Ca^{2+} signal were agonist-specific (Schöfl et al 1991, Sanchez-Bueno et al 1993). It therefore appears, that at least in hepatocytes the intracellular Ca^{2+} signal is amplitude and frequency encoded, which might allow for differential activation of distinct Ca^{2+}-controlled processes depending on co-activation of other second messenger pathways and on the temporal pattern of agonist stimulation (Schöfl et al 1994).

Prank: You addressed the question of amplitudes. Another issue is the relevance of the precision of Ca^{2+} spiking for the regulation of the intracellular decoding machinery.

Fields: Have you thought of applying this kind of analysis to the next downstream event, such as Ca^{2+}-activated kinase or transcription factors?

Prank: Yes, we are working on correlating the measures for the coding behaviour to downstream biological responses that are dependent on the temporal pattern of Ca^{2+}, such as the activation of the Ca^{2+}/calmodulin-dependent kinase II (CaMKII). The methods used in this study are of course applicable to investigate the information flow to other downstream events, such as activation of gene expression.

Sejnowski: To what extent could the same approach be used to look at frequency dependence in other signalling systems? For example, the differentiation of spinal cord cells depends on the frequency of the Ca^{2+} spikes. In the developing spinal cord, depending on the frequency, progenitor cells can become interneurons or excitatory cells (Gu & Spitzer 1995). There must be some way that the frequency affects the differentiation: the actual genes that are turned on or off. How would you incorporate this into your model?

Prank: Whether genes are turned on or off is a binary decision. The reconstruction method used in our study to explore the dynamics of transmembrane signalling is based on dynamic input pattern and dynamic output pattern. Thus, it is conceivable that the input can be a Ca^{2+} spike train and the output the time course of a Ca^{2+}-dependent process, such as the phosphorylation of an enzyme, a target protein, the dynamics of secretion, or the dynamics of gene expression. However, the reconstruction method requires dynamic input and output.

Sejnowski: Let me try to be more explicit. There is one point of view, which is that you are trying to preserve as much of the information coming in as possible. This is the traditional view of information theory, and is the approach that you have taken. There is a different view, which is that the receiver is trying to detect a stereotyped or simple signal buried in a lot of noise. Therefore, the amount of information that it is receiving is not the entire bandwidth that is coming through, but a small fraction of it. In this case this kind of analysis might be deceiving.

Prank: This is an important point: what is the receiver trying to detect? To answer this question, one might relate the measures for the quality of coding, such as the coding fraction or information rate, to the biological responses downstream from the Ca^{2+}-signal. You might then address the question, from one step to the other of the intracellular encoding and decoding machinery, of how much of the bandwidth from the stimulus to Ca^{2+} signal is preserved in the final biological response, such as the information coded in the levels of phosphorylated target protein, the pattern of gene expression and other responses.

Berridge: Such experiments have been done on T cells, where differential gene activation was recorded when Ca^{2+} spikes were applied at different frequencies. It comes back to the question Sydney Brenner asked earlier: how much information can you transmit using one signal? It seems that the cell can use Ca^{2+} to regulate a number of different processes simply by varying both amplitude and frequency.

Laughlin: There are cases where reconstruction techniques have been misapplied to neurons because random stimuli are inappropriate. Many neurons are rather poor at monitoring random stimuli and reject white noise, because the nervous system has evolved to detect and process patterns that are, by definition, non-random. Consequently such a neuron's response to white noise is so weak that the reconstruction is pitiful and tells us virtually nothing about the neuron's function. It is remarkable, therefore, that the reconstruction of random inputs obtained by Klaus Prank is quite good. The reconstruction tells us that detailed information on how the hormone concentration is changing, minute-by-minute, is available to mechanisms within the cell. We learn that the second messenger system is not averaging the signal over such long time intervals that all temporal fluctuations are removed. Whether or not these details are used by downstream mechanisms is a matter for further investigation, but we now know that they are there. That is an advance.

References

Brabant G, Prank K, Schöfl C 1992 Pulsatile patterns in hormone secretion. Trends Endocrinol Metab 3:183–190

Gu X, Spitzer NC 1995 Distinct aspects of neuronal differentiation encoded by frequency of spontaneous Ca^{2+} transients. Nature 375:784–787

Sanchez-Bueno A, Marrero I, Cobbold PH 1993 Different modulatory effects of elevated cyclic AMP on cytosolic Ca^{2+} spikes induced by phenylephrine or vasopressin in single rat hepatocytes. Biochem J 291:163–168

Schöfl C, Sanchez-Bueno A, Brabant G, Cobbold PH, Cuthbertson KSR 1991 Frequency and amplitude enhancement of calcium transients by cyclic AMP in hepatocytes. Biochem J 273:799–802

Schöfl C, Brabant G, Hesch RD, von zur Mühlen A, Cobbold PH, Cuthbertson KS 1993 Temporal patterns of alpha 1-receptor stimulation regulate amplitude and frequency of calcium transients. Am J Physiol 265:C1030–1036

Schöfl C, Prank K, Brabant G 1994 Mechanisms of cellular information processing. Trends Endocrinol Metab 5:53–59

Woods NM, Cuthbertson KSR, Cobbold PH 1986 Repetitive transient rises in cytoplasmic free calcium in hormone stimulated hepatocytes. Nature 319:600–602

From genes to whole organs: connecting biochemistry to physiology

Denis Noble

University Laboratory of Physiology, University of Oxford, Parks Road, Oxford OX1 3PT, UK

Abstract. The successful analysis of physiological processes requires quantitative understanding of the functional interactions between the key components of cells, organs and systems, and how these interactions change in disease states. This information does not reside in the genome, or even in the individual proteins that genes code for. There is therefore no alternative to copying nature and computing these interactions to determine the logic of healthy and diseased states. The rapid growth in biological databases, models of cells, tissues and organs, and in computing power has made it possible to explore functionality all the way from the level of genes to whole organs and systems. Examples are given of genetic modifications of the Na^+ channel protein in the heart that predispose people to ventricular fibrillation, and of multiple target therapy in drug development. Complexity in biological systems also arises from tissue and organ geometry. This is illustrated using modelling of the whole heart.

2001 Complexity in biological information processing. Wiley, Chichester (Novartis Foundation Symposium 239) p 111–128

Beyond the genome: the role of modelling

The amount of biological data generated over the past decade by new technologies has completely overwhelmed our ability to understand it. Genomics has provided drug discoverers with a massive 'parts catalogue' for the human body, while proteomics seeks to define these individual 'parts' and their structure in detail. But there is as yet no 'user's guide' describing how these parts interact to sustain life or cause disease. In many cases, the cellular, organ and system functions are unknown, though clues often come from homology in the gene sequences. Moreover, even when we understand function at the protein level, successful physiological and pharmaceutical intervention depends on knowing how a protein behaves in context, as it interacts with the rest of the relevant cellular machinery to generate function at a higher level. Without this integrative knowledge, we may not even know in which disease states the proteins are relevant, and we will certainly encounter side- and counterintuitive-effects that are unpredictable from molecular information alone.

Inspecting genome databases alone will not get us very far in addressing these problems. The reason is simple. Genes code for protein sequences. They do not explicitly code for the interactions between proteins and other cell molecules and organelles that generate function. Nor do they indicate which proteins are on the critical path for supporting cell and organelle function in health and disease. Much of the logic of the interactions in living systems is implicit. Wherever possible, nature leaves much of the detail to be determined by the chemical properties of the molecules themselves and to the exceedingly complex way in which these properties have been exploited during evolution. Thus, nothing in the genome codes for the properties of water but these properties, like many other naturally occurring physicochemical properties, are essential to life as we know it. It is as though the function of the genetic code, viewed as a program, is to build the components of a computer, which then self-assembles to run programs about which the genetic code knows nothing. At a previous Novartis Foundation Symposium, Sydney Brenner (1998) expressed this very effectively when he wrote: 'Genes can only specify the properties of the proteins they code for, and any integrative properties of the system must be "computed" by their interactions'. Brenner meant not only that biological systems themselves 'compute' these interactions but also that in order to understand them *we* need to compute them, and he concluded, 'this provides a framework for analysis by simulation'.

Computer models must be used when the complexity of a system is too great to grasp intuitively. This has increasingly become the case in biology (Bailey 1999). Proteins must interact with many other proteins depending on time and place, individual molecules may participate in multiple pathways, and the background against which protein function is expressed can change dynamically with sex, age and disease. Models are used to hypothesize new approaches, and to identify where gaps in knowledge exist. A modeller can take a set of existing data and determine whether those data are sufficient to generate the output of the system under study. If they are not, the modeller can then suggest specific directions for further study as well as predictions about possible results. This iterative interaction between modelling and experimentation is essential for success. Conducting 'experiments' in a virtual environment where the possible impact of different conditions can be tested systematically allows the researcher to select the best overall design principles in advance of real life studies. Modelling, therefore, is an essential tool of analysis.

It is also of practical importance since the computational modelling of biological systems can add significant value in the discovery and development of new therapeutic agents (Noble & Colatsky 2000). One can understand protein function in context, identify and validate new drug targets against the background in which the function of that target is expressed, and understand the impact of clinical variables on drug action in ways that cannot be adequately represented in even the most complex animal models. Computer modelling of cells and organs can help the researcher conduct virtual genetic studies in which cellular components are 'knocked-out', 'knocked-in' or modified by genetic mutation, and then to use this information to design new drugs or to carry out a more advanced research plan. Given the complexity of many diseases, models also provide an additional advantage by allowing us to define the optimal therapeutic profile of a new drug prior to chemical synthesis. For example, the researcher can explore in a rational and systematic way whether the most effective treatment is a drug that acts specifically on a single target or one that acts at multiple targets, and in what relative proportion these additional activities should occur. Finally, one can prospectively investigate issues of clinical safety and efficacy using models developed to answer questions about toxicology and pharmacodynamics.

Non-linear effects of a genetic mutation

In this article I will illustrate some of these general principles using computer modelling of the heart. The first example is that of genetic mutations in the Na^+ channel protein that can cause ventricular fibrillation, a life-threatening event in which the heart beats in a highly asynchronous fashion and ceases to act as an efficient pump. Several such mutations are known. One of these is a missense mutation that has a well-characterized effect on the function of the Na^+ channel protein: the inactivation curve is shifted by a few millivolts in the depolarizing direction (Chen et al 1998). By itself, this information does not enable us to make any predictions about the effect on the heart. However, by inserting this information into the highly complex models that have now been developed for cardiac cells, we can make some very interesting predictions.

One of the effects of this mutation is to shift the voltage dependence of Na^+ channel inactivation. Figure 1 shows the result of inserting 12 and 18 mV shifts of the inactivation curve into a ventricular cell model (Noble et al 1998). We chose these two values first because they fall within the range of the experimental data on Na^+ channels expressed in oocytes and, second, because they illustrate how highly non-linear the overall response may be. In fact, a 12 mV shift (curve b) has only a small effect on the computed action potential. The repolarization phase is prolonged by an amount that would be too small to have any significant effect at the level of the whole organ. By contrast, adding a further 6 mV of voltage shift gives a qualitatively different response (curve c). The late phase of repolarization is interrupted by a series of after-depolarizations (EADs), leading to massive prolongation of the action potential. This kind of extreme non-linearity,

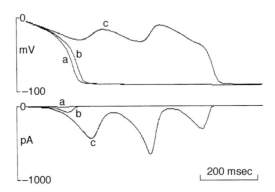

FIG. 1.   Reconstruction of the arrhythmogenic effects of *SCN5A* gene mutation. Action potential repolarization (top) and Na⁺ current (bottom) are shown in three different conditions: (a) normal Na⁺ channel; (b) model of *SCN5A* gene mutation, expressed by a moderate positive shift in Na⁺ channel inactivation curve; and (c) as before, but with a more severe shift of inactivation. Reprinted with permission of The Physiological Society from Noble & Noble (1999).

involving a qualitative change in behaviour frequently occurs in complex systems. Moreover, the sudden occurrence of EADs is observed in several other pathological conditions that predispose the heart to fatal arrhythmia.

Similar results have been obtained using mutations of the Na⁺ channel that underlie the long QT syndrome (Clancy & Rudy 1999).

Multiple changes in gene expression levels: the example of congestive heart failure

Congestive heart failure is an example in which the molecular mechanisms of the EADs are very different but the end result is very similar. This illustrates another property of complex systems, which is that very similar outcomes may be generated by very different underlying molecular processes. In this case, the early after-depolarizations arise from changes in the expression levels of several membrane transporters other than the Na⁺ channel. Figure 2 summarizes these changes. Two surface membrane potassium channels ($I_{to,1}$ and i_{K1}) have their expression levels reduced, which prolongs the action potential and predisposes the cells to the generation of EADs. The sarcoplasmic reticulum Ca^{2+} pump (SERCA, referred to as I_{up} in this diagram) is also reduced in activity. One transporter, the Na^+/Ca^{2+} exchanger, is up-regulated.

Figure 3 shows the results of incorporating this particular EAD mechanism into a model of the whole ventricles. In place of the ordered spread of excitation following each sinus beat (as in A) we observe continuously re-entrant waves of

Molecular basis of CHF

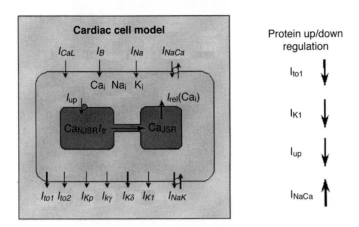

FIG. 2.   Some of the key transporters involved in cardiac cell activity together with their changed levels of expression in congestive heart failure (CHF). Based on Winslow et al (1998).

excitation that meander in a complex way around the heart (B). This form of re-entrant arrhythmia generates a characteristic triangular ECG whose amplitude waxes and wanes slowly. This is the arrhythmia characterized as Torsades de Pointes, and it is well reconstructed by the computer model.

There is another important difference to note here. In the case of the Na^+ channel mutation, a *single* molecular event is responsible for the pathology, whereas in the case of congestive heart failure, multiple molecular changes are responsible. It is likely that the latter is the more frequent case than the former. This has extremely important consequences for the development of therapeutic approaches. The standard 'classical' pharmaceutical approach to arrhythmia therapy, for example, has been to develop compounds that target a particular receptor or transporter. In fact, the major system of classification of anti-arrhythmic drugs (the Vaughan-Williams classification) is based on identifying the channel mechanism on which the drug acts. Ideally, it has been thought, we should look for 'pure' drugs that have a single action with few 'side effects'. This approach has been a spectacular and expensive failure. Clinical trials of many anti-arrhythmic compounds developed using this approach have been either disappointing or disastrous (see e.g. CAST Investigators 1989)—so much so that many pharmaceutical companies no longer have a major drug discovery programme in this area.

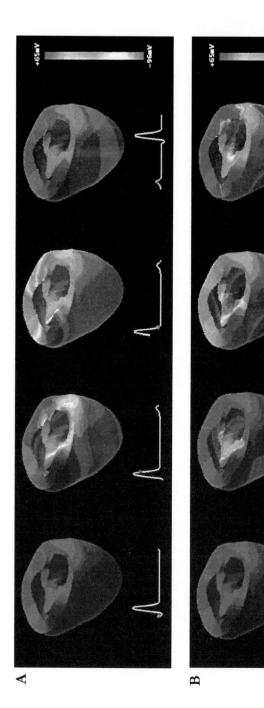
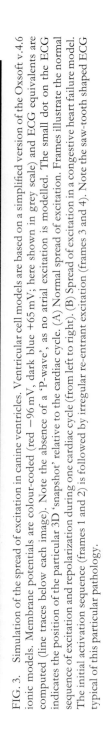

FIG. 3. Simulation of the spread of excitation in canine ventricles. Ventricular cell models are based on a simplified version of the Oxsoft v.4.6 ionic models. Membrane potentials are colour-coded (red −96 mV, dark blue +65 mV; here shown in grey scale) and ECG equivalents are computed (line traces below each image). Note the absence of a 'P-wave', as no atrial excitation is modelled. The small dot on the ECG indicates the position of the particular 3D 'snapshot' relative to the cardiac cycle. (A) Normal spread of excitation. Frames illustrate the normal sequence of excitation and repolarization during one cardiac cycle (from left to right). (B) Spread of excitation in a congestive heart failure model. The initial activation sequence (frames 1 and 2) is followed by irregular re-entrant excitation (frames 3 and 4). Note the saw-tooth shaped ECG typical of this particular pathology.

Multiple target therapy: example of an anti-arrhythmic compound

This approach is clearly flawed. In fact, the failure is not surprising. In complex systems in which many proteins interact there is little reason to expect that intervening on a *single* molecular mechanism will be effective or that the outcome will be easily predictable. Instead of requiring a solo performance, we should perhaps be expecting our therapies to play a more complex performance, tailored to the particular pathology and its profile of changes in expression levels of the proteins involved.

A good example of this approach is the compound BRL-32872 (SmithKline-Beecham), which was for a period a lead compound in anti-arrhythmia therapy. The objective of this drug discovery program was to obtain the therapeutic effects of action potential prolongation via K^+ channel blockade, without triggering Torsades de Pointes arrhythmias (Bril et al 1996). In the case of K^+ channel blockers the mechanism of these early after-depolarizations is that the L-type Ca^{2+} channels generate a strong 'window' current in the range of potentials at which the rapid phase of repolarization begins. This is a critical phase of repolarization where an imbalance of ionic currents can switch the process over from smooth repolarization to re-excitation. The logical aim therefore should be to combine K channel blockade with just enough L-type Ca^{2+} channel block to ensure that repolarization continues to occur smoothly even in a prolonged action potential. BRL-32872 succeeded in achieving action potential prolongation without triggering Torsades de Pointes in animal experiments.

Figure 4 shows that the cardiac action potential models are fully capable of reproducing this logic. 90% block of i_K clearly triggers EADs, whereas adding just 20% block of L-type Ca^{2+} channels completely abolishes the EADs, while still giving nearly all the action potential prolongation.

Although BRL-32872 was later dropped as a lead compound because of metabolic side effects, the approach is clearly valid, and other compounds with this multi-receptor action profile are now under investigation (Nadler et al 1998). One use of the models is to trawl through the many other combinations of two and three site actions to determine which would be therapeutic and which could be expected to be dangerous.

Complexity arising from tissue and organ geometry

The examples of complexity I have given so far are those in which multiple molecular interactions occur which are difficult to understand or predict without integrative experimental and computational analysis. In the final part of this article I will emphasise another source of complexity in biological systems: that of complex tissue and organ geometry (Hunter et al 1997, 1998, Hunter & Smaill 1989, Kohl et al 2000).

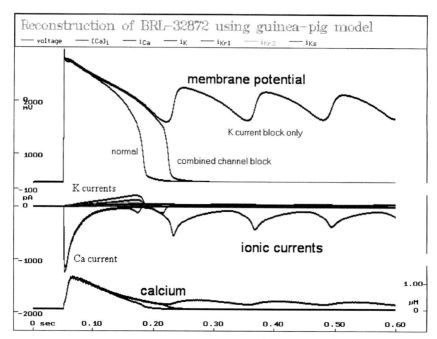

FIG. 4.   Reconstruction of the action of a multiple target compound, BRL-32872. Top traces: computed membrane potentials in normal situation, with K^+ channel block only and with combined K^+ and Ca^{2+} channel block. Middle traces: computed membrane currents. Lower traces: computed Ca^{2+} transients. At therapeutic concentrations this compound produces 90% block of the K^+ channel, i_{Kr}, which would be expected to prolong the action potential but at the price of inducing early after-depolarizations of the kind that underlie Torsades de Pointes arhhythmias. At the cell level these are represented by the repetitive after-depolarizations computed here. In fact, BRL-32872 also produces 20% block of the Ca^{2+} channel, i_{Ca}, which is sufficient to abolish the early after-depolarizations while still generating the action potential prolongation. From Noble & Colatsky (2000).

The ventricular geometry used in the model shown in Fig. 3 is based on measurements of the epicardial and endocardial surfaces of both ventricles of a canine heart, fitted with a finite element model to an accuracy of about 0.5 mm (Nielsen et al 1991). In addition to general geometry, the fibrous-sheet structure of ventricular myocardium (LeGrice et al 1997) is also represented by finite element model parameters, yielding a continuous description of fibre and sheet orientations throughout the myocardium. Fibre direction and sheet orientation determine passive and active mechanical properties, as well as key electrical characteristics, including patterns of conduction (see Fig. 5). Active contraction is triggered in the model via excitation-contraction coupling. The underlying electrical properties of cells can be defined to represent any of the single cell models.

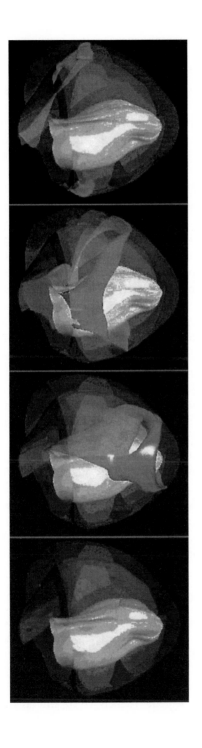

FIG. 5. Spread of the electrical activation wavefront in an anatomically detailed cardiac model. Earliest activation occurs at the left ventricular endocardial surface near the apex (left). Activation then spreads in endocardial-to-epicardial direction (outwards) and from apex towards the base of the heart (upwards, middle frames). The activation sequence is strongly influenced by the fibrous-sheet architecture of the myocardium, as illustrated by the non-uniform transmission of excitation. Red, activation wavefront; blue, endocardial surface (shown here as grey scale).

Furthermore, the first six generations of the coronary tree, starting with the large epicardial vessels and ending with vessels of the order of $100\,\mu m$ diameter, are represented discretely (see Fig. 6) (Kassab et al 1993, Smith 1999). A black box model of the capillary bed is used to connect arterial and venous vessels in the model.

Solving this anatomically representative, electro-mechanical model of the heart requires powerful super-computational equipment. I will illustrate two examples of such studies, performed on a Silicon Graphics 16-processor (R10000) shared memory Power Challenge.

In the first example, the spread of ventricular activation is modelled (Fig. 5). In this case, the membrane potential is represented by a FitzHugh–Nagumo type model. Excitation is initiated by a stimulus point on the left ventricular endocardium near the apex (earliest breakthrough point). Spread of the activation wavefront is heavily influenced by cardiac ultrastructure, with preferential conduction along the fibre-sheet axes referred to above (Sands 1998).

The second example combines contraction and coronary tree architecture in one model that allows simulation of changes in intra-luminal coronary pressure during the cardiac cycle. The coronary tree moves with the cardiac tissue into which it is embedded and the transmural pressure acting on the vessels is calculated from the difference between fluid pressure in the coronaries and external stress. This pressure is shown with the deforming coronary vessel tree in Fig. 6 (Smith 1999).

Thus, complex electromechanical models of ventricular anatomy and function allow one to describe coronary perfusion during the cardiac cycle. By linking this to models of cell metabolism (Ch'en et al 1998) and electromechanical function, the whole sequence from a simulated disturbance in coronary blood supply to depression in ventricular pressure development may be computed. This creates an immense potential, not only for biomedical research but also for clinical applications, including patient-specific modelling of therapeutic interventions. This approach could, for example, be used for the prediction of optimal coronary bypass procedures, as modelling of a patient's cardiac anatomy is feasible on the basis of nuclear magnetic resonance data (Young et al 1996a,b) and 3D coronary angiography can provide data on coronary tree architecture.

Conclusion: from genome to proteome to physiome

The human genome is a vast database of information containing in the order of 50 000 to 150 000 genes. Each of these is used to determine the amino acid sequence of a particular protein. The complete sequence and structure of the proteins is sometimes referred to as the proteome. Understanding how the information in the genome is used to create the proteome is a major challenge, first because we need to identify all the genes (which we are still far from doing)

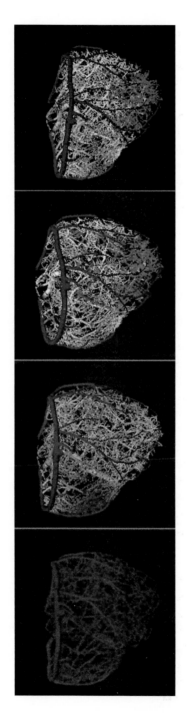

FIG. 6. Flow calculations coupled to the deforming myocardium. The colour coding represents transmural pressure acting on the coronary vessels from the myocardial stress (dark blue, zero pressure; red, peak pressure; shown here as grey scale). The deformation states are (from left to right) zero pressure, end-diastole, early systole, and late systole.

and, second, because predicting three-dimensional structure and chemical function from the amino acid sequences of the corresponding proteins is very difficult (Onuchic et al 1997). But even these major challenges pale in significance when we consider the complexity of the next stage: understanding the interactions of tens of thousands of different proteins as they generate functionality at all levels through cells to organs and systems. This is the task of quantitative analysis of physiological function, which in its entirety is sometimes now called the physiome (Bassingthwaighte 1995). Computational modelling will play an increasingly important role in all these stages of unravelling the way in which the information contained in the genome is 'read' to create living systems. We will be able to say that we have really read 'The Book of Life' when we have succeeded in going all the way from the genome, through the proteome to the physiome.

References

Bailey JE 1999 Lessons from metabolic engineering for functional genomics and drug discovery. Nat Biotechnol 17:616–618

Bassingthwaighte JB 1995 Towards modeling the human physionome. In: Sideman S, Beyar R (eds) Molecular and subcellular cardiology: effects on structure and function. Plenum, New York, p 331–339

Brenner S 1998 Biological computation. In: The limits of reductionism in biology. Wiley, Chichester (Novartis Found Symp 213) p 106–116

Bril A, Gout B, Bonhomme M et al 1996 Combined potassium and calcium channel blocking activities as a basis for antiarrhythmic efficacy with low proarrhythmic risk: experimental profile of BRL-32872. J Pharmacol Exp Ther 276:637–646

CAST Investigators 1989 Preliminary report: effect of encainide and flecainide on mortality in a randomized trial of arrhythmia suppression after myocardial infarction. N Eng J Med 321:406–412

Ch'en FF, Vaughan-Jones RD, Clarke K, Noble D 1998 Modelling myocardial ischaemia and reperfusion. Prog Biophys Mol Biol 69:515–538

Chen Q, Kirsch GE, Zhang D et al 1998 Genetic basis and molecular mechanism for idiopathic ventricular fibrillation. Nature 392:293–296

Clancy CE, Rudy Y 1999 Linking a genetic defect to its cellular phenotype in a cardiac arrhythmia. Nature 400:566–569

Hunter PJ, Smaill BH 1989 The analysis of cardiac function: a continuum approach. Prog Biophys Mol Biol 52:101–164

Hunter PJ, Nash MP, Sands GB 1997 Computational electromechanics of the heart. In: Panfilov A, Holden A (eds) Computational biology of the heart. Wiley, Chichester, p 345–407

Hunter PJ, McCulloch AD, ter Keurs HEDJ 1998 Modelling the mechanical properties of cardiac muscle. Prog Biophys Mol Biol 69:289–331

Kassab GS, Rider CA, Tang NJ, Fung YC 1993 Morphometry of pig coronary arterial trees. Am J Physiol 265:H350–H365

Kohl P, Noble D, Winslow RL, Hunter P 2000 Computational modelling of biological systems: tools and visions. Phil Trans R Soc Lond A Phys Sci 358:579–610

LeGrice IJ, Hunter PJ, Smaill BH 1997 Laminar structure of the heart: a mathematical model. Am J Physiol 272:H2466–H2476

Nadler G, Faivre JF, Forest MC et al 1998 Synthesis, electrophysiological properties and analysis of structural requirements of a novel class of antiarrhythmic agents with potassium and calcium channel blocking properties. Bioorg Med Chem 6:1993–2011

Nielsen PMF, LeGrice IJ, Smaill BH, Hunter PJ 1991 A mathematical model of the geometry and fibrous structure of the heart. Am J Physiol 29:H1365–H1378

Noble D, Colatsky TJ 2000 A return to rational drug discovery: computer-based models of cells, organs and systems in drug target identification. Emerg Therap Targets 4:39–49

Noble D, Noble PJ 1999 Reconstruction of cellular mechanisms of genetically-based arrhythmias. J Physiol 518:2–3P

Noble D, Varghese A, Kohl P, Noble P 1998 Improved guinea-pig ventricular cell model incorporating a diadic space, i_{Kr} and i_{Ks}, and length- and tension-dependent processes. Can J Cardiol 14:123–134

Onuchic JN, Luthey-Schulten Z, Wolynes PG 1997 Theory of protein folding: the energy landscape perspective. Annu Rev Phys Chem 48:545–600

Ruben PC, Vilin YY, Fujimoto E 2000 Molecular basis of slow inactivation in sodium channels. J Physiol, in press

Sands GB 1998 Mathematical model of ventricular activation in an anatomically accurate deforming heart. PhD thesis, Department of Engineering Science, University of Auckland, Auckland, New Zealand

Smith NP 1999 Coronary flow mechanics: an anatomically accurate mathematical model of coronary blood flow coupled to cardiac contraction. PhD thesis, Department of Engineering Science, University of Auckland, Auckland, New Zealand

Winslow RL, Rice J, Jafri S 1998 Modeling the cellular basis of altered excitation-contraction coupling in heart failure. Prog Biophys Mol Biol 69:497–514

Young AA, Orr R, Smaill BH, Dell'Italia LJ 1996a Three-dimensional changes in left and right ventricular geometry in chronic mitral regurgitation. Am J Physiol 271:H2689–H2700

Young AA, Fayad FA, Axel L 1996b Right ventricular midwall surface motion and deformation using magnetic resonance tagging. Am J Physiol 271:H2677–H2688

DISCUSSION

Segel: Have you included the fluids in your model of the heart?

Noble: We are computing the pressure changes of the blood inside the heart (see for example Fig. 6 in my paper), but you will notice two things missing in our model. First of all there are no valves and, second, the atrium and sinus node are missing. These are jobs in progress. We are in the process of putting together the atrial and sinus node anatomy; this is more complicated than the ventricular anatomy. We think we have the right models for the cellular behaviour in the sinus node and atrium, and we have models of the valves.

Segel: Do you plan to employ Peskin's immersed boundary approach to the fluids (McQueen & Peskin 1997, 2000), or something different?

Noble: I'm not the right person to ask. The team tackling the flow side of this project are our collaborators in Auckland, New Zealand. I can guess what their answer would be, though: our equations are going to be much the same, because the fundamental equations of flow are fairly obvious. There is one difficulty with modelling complicated systems, particularly when you are going all the way from biochemical and genetic changes up to the whole organ level, which is that the number of forms of expertise involved in doing this is quite phenomenal. My latest estimate of the total team involved in making sure that this happens around the world is 80 scientists, of which 20 are working in Oxford, 20 in New Zealand, 20 in Johns Hopkins and 20 in Princeton. No one of us knows the whole story.

Segel: I have a question about your general philosophy. Let us take channels as an example. There are many different channel types: do you start with most of the known channels and then remove them if they don't do anything interesting, or do you build up from a simple Hodgkin–Huxley model, adding more channels as you need them to obtain significant observed behaviour?

Noble: We have multiple models of nearly all the channels. Take the Na^+ channel. What I used in my presentation was the simple Hodgkin–Huxley type of model, but we also have multistate models, running up to the full set of a dozen or so states that some people have postulated. For the details of that genetic mutation, not the shift of the inactivation curve (Fig. 1 in my paper) but the other features—some of the detailed kinetics of the changes—we are unfortunately going to have to graduate to the multistate models. We are in the process at the moment of trying to fit experimental data from the gene mutation information on the kinetics obtained by Peter Ruben and his colleagues in the USA to the multistate models (Ruben et al 2000). The problem we encounter is that it is not very well defined. There is a major difficulty with multistate models of channels, because by and large we don't have enough data from the kinetic information to give unique fits. We desperately need better ways of achieving this, which is taking time. For other mechanisms, such as K^+ channels, again we have different levels of models. You might be wondering why we don't put all the data in and go to the most complicated of each of the transporter models. Why cut down? The obvious reason for cutting down has to do with the volume of data and computability. Even with the massive computing power of some of the parallel machines we have access to, we need nevertheless to be economical on total computing time. The other reason is one of more general interest to this symposium: we can get fooled by the complexity of what we have done. We then have great difficulty understanding precisely what has emerged. Let me give one example. Ischaemic arrhythmias are thought to be generated by the fact that the Ca^{2+} oscillator acts on the Na^+/Ca^{2+} exchanger to produce current flowing into the cell through Na^+ trying to push the Ca^{2+} out: this current generates the depolarization, which is why it generates arrhythmia. The classic pharmaceutical approach to this problem would be to use a drug that blocks that particular transporter to try to control the arrhythmia. What we discovered when we ran this computation was that this doesn't work! In fact, it makes things worse. This gave us a very interesting counterintuitive hint. If blocking it makes the situation much worse, why not go in the other direction and up-regulate this mechanism?

That works (Ch'en et al 1998). There are already transgenic mice with up-regulated Na^+/Ca^{2+} exchangers, so there is a lovely experimental test of this. We will have to put the hearts of these mice into ischaemia and see whether they resist Na^+/Ca^{2+} overload much better and thus don't go into cardiac arrhythmia. The point I am making is as follows. Coming back to the question of complexity, we have to make sure that we don't build models that are so complex that we can't understand them. We need to unravel this counterintuitive result. In this case, we asked a research student, who is a mathematician, to take on as his PhD project the task of building a simpler model of the subset of components that we think is essential for that process and then work out mathematically why it is that up-regulation of the Na^+/Ca^{2+} exchanger would be predicted to be beneficial. This does two things. First of all it gives you better understanding. We sometimes have to go into a much simpler system, retaining the main features of the complex system, in order to get the understanding, and then we can take that back into the more complex full scale system and start to see whether our simple system is a good model mathematically of the computational detail at a higher level. So, we have multiple models of the various channels and transporters. We do not apologise for this because we think they are needed. This relates to another matter that I think we should address in the general discussion later on, which is what is the philosophy of modelling complex systems? The second thing is that to understand what we have done, we have often got to simplify it.

Segel: Do you ever throw out channels that don't seem to be doing anything?

Noble: Sydney Brenner said something earlier about it not being cost effective for an organism to knock things out that were no longer used; that it might be better to leave them there. The heart has a beautiful way of doing this. The pacemaker current is present in all parts of the heart, and in the embryo it works in all parts of the heart. Embryonic ventricular cells will beat away even if they are isolated. This was first demonstrated by William Harvey, a few hundred years ago. He took the heart of an eel, put some of his spittle on his hand and proceeded to chop the heart up into tiny bits. Every single bit went on beating. This means that the eel heart is like the embryonic human heart: it has functional pacemaker current in all regions. But the mammalian heart doesn't switch this system off in development, it just changes the voltage dependence to take it out of the physiological range (except of course in the adult pacemaker regions). When we first did experiments on atrial cells, we hyperpolarized them beyond the normal physiological range, well below the resting potential, and lo and behold we found our dear friend the pacemaker current. It is still there. So far, in every single cardiac cell that has been investigated, if you do it with fresh enough cells, the pacemaker current is there. However, in the modelling, we just knock out everything that we think is not functionally relevant to the cell.

Sejnowski: Are there conditions when it does hyperpolarize under abnormal conditions?

Noble: That is a good question. Can a cardiac cell ever go below −95 mV? My answer would be no, because that is the K$^+$ equilibrium potential. There is therefore no current that could drive it beyond that voltage. As far as I can see, unless you come along with a great big defibrillator — and who knows how defibrillators work, incidentally — I don't think cells are ever hyperpolarized beyond −95 mV. So what is a channel that only activates at −120 mV doing there? My theory is that shifting the activation curve is the easiest way to get rid of the channel, i.e. to render it non-functional.

Laughlin: It is very impressive that this system, which doesn't have a huge number of molecular components, is complicated and demanding to model.

Noble: We are modelling up to about 100 molecular components in a single cell, but we don't always include all of these in all simulations.

Laughlin: You cannot possibly understand how this system works without this model.

Noble: I'd also say that you can't understand it without an understanding of the physiology. I think this is an interaction between intuitive physiological understanding and computational understanding.

Laughlin: The critical question for those of us who don't work on the heart is, how typical are the interactions that require this complex modelling of other systems?

Noble: You are raising not just a question of typicality, but also one of validation. These are connected, because if you can validate the model you can decide how typical the behaviour is. At the cellular level, for reconstruction of basic electrophysiology, that iteration between experiment and theory has gone on for many years (Noble & Rudy 2001). I have 40 years' experience of doing this: I first modelled a single cardiac cell way back in 1960, when the computer we used was a huge valve machine filling a room. I am not saying that there is not room for immense improvement — there is — but I think for many purposes, particularly insertion into whole organ models and linking biochemistry and physiology, there have been enough rounds of iteration to feel some degree of confidence. But as we go on to the newer parts of the modelling, we feel much less confident. For the metabolism modelling, we compared results with those obtained by NMR studies on whole hearts. We got reasonably good results in terms of the changes in basic metabolites during ischaemia. Nevertheless, on that kind of modelling we still need to add a lot more detail. Can we get whole heart validation? Thank goodness there are a few whole organ physiologists left! Very fortunately, one of these is David Paterson, working in the physiology laboratory in Oxford, who still does experiments on whole hearts, plunging a vast number of electrodes into dog and pig hearts. We are collaborating with him, trying to check whether our simulations of the whole heart in terms of spread of excitation and so on, are validated reasonably well by experimental data. So, I think that where we have gone through a lot of iteration, the results are reasonably typical. But there is a lot of fringe work, particularly in the newer modelling, where I think we have a long way to go before we can say that the system is validated.

Laughlin: I was actually using typicality in a difference sense: whether the bits and pieces that make this system complicated are the same bits and pieces that are found in other systems, which might make them equally complicated.

Noble: What is going to happen when we go on to model the lung, for example? We are on the way to developing a lung structure. We are in the process of using our expertise with modelling the electrophysiology of cardiac cells to see whether we can do the same for tracheal smooth muscle. The aim would be to model what must go wrong in asthma or the effects of pollutants on breathing. To answer your question, we have got quite a long way with modelling tracheal cells using our heart expertise. Of course, the big difference is that in this case the Ca^{2+} oscillator is driven by an $InsP_3$ receptor rather than a ryanodine receptor. So far as we know, quite a bit of the expertise with modelling channels can be carried over. The basic behaviour of the Na^+/Ca^{2+} exchanger, for example, is the same.

Brenner: There is now a rich resource of human mutations, which we can correlate with the physiology of the organism. Using this information, together with our ability to make mutations in the isolated components, to look at molecular function, will be helpful in providing the bridge. The use of specific drugs that allow us to dissect the physiology is also going to be very important.

Sejnowski: A problem with many drugs is that they are not specific.

Brenner: They are not specific, but none the less they have allowed tremendous insights.

Noble: It might sometimes be a good thing if the drugs are not specific. Let me give an example. The standard pharmaceutical approach to arrhythmia is to identify the channel that is the immediate cause and to try to develop a pure blocker for that channel. Between 1989–2000 this led to six catastrophic clinical trials on anti-arrhythmic drugs, and now practically all the main pharmaceutical companies are out of the game of developing drugs for cardiac arrhythmia, despite the fact that this is a US$50 billion market. We found that if we add a 20% block of the Ca^{2+} channel to a K^+ channel blocker, we get a superb 'virtual' compound (see Fig. 4 in my paper). This is a good example of where modelling can tell you which combination of receptors you should go for, and even in which proportion.

Sejnowski: I know that there is another approach to arrhythmia based on a top–down model, involving bifurcation theory. To what extent does this approach map on to what you have done?

Noble: In the case of the Ca^{2+} oscillator, that is precisely what we did: we stripped down to a simple set of equations which would give the minimum conditions for

Ca^{2+} oscillation but be capable of treatment by bifurcation theory, chaos theory and so on. This is producing some valuable insights. The short answer to your question is that I think that we do need to map from the simpler models onto the complicated, computationally intense models, partly to get understanding but partly also because we may find that for many purposes in whole organ simulation the simple models will do. This relates to a point I would like to raise in the general discussion about levels of modelling.

Brenner: I think it is important to do it this way because you want to intervene. It is very hard to intervene on bifurcation equations. That indeed would be a great drug!

References

Ch'en FF, Vaughan-Jones RD, Clarke K, Noble D 1998 Modelling myocardial ischaemia and reperfusion. Prog Biophys Mol Biol 69:515–538

McQueen DM, Peskin CS 1997 Shared-memory parallel vector implementation of the immersed boundary method for the computation of blood flow in the beating mammalian heart. J Supercomputing 11:213–236

McQueen DM, Peskin CS 2000 A three-dimensional computer model of the human heart for studying cardiac fluid dynamics. Computer Graphics 34:56–60

Noble D, Rudy Y 2001 Models of cardiac ventricular action potentials: iterative interaction between experiment and simulation. Phil Trans Roy Soc B Biol Sci, in press

Ruben PC, Vilin YY, Fujimoto E 2000 Molecular basis of slow inactivation in sodium channels. J Physiol, in press

Development of high-throughput tools to unravel the complexity of gene expression patterns in the mammalian brain

U. Herzig, C. Cadenas, F. Sieckmann*, W. Sierralta†, C. Thaller‡, A. Visel and G. Eichele[1]

*Max-Planck-Institute of Experimental Endocrinology, Feodor-Lynen-Strasse 7, 30625 Hannover, Germany, *Frank Sieckmann, Hörsterholz 1d, D-44879 Bochum, Germany, †Laboratorio de Ultraestructuras, Division Nutricion Humana, INTA-Universidad de Chile, Avda. Jose P. Alessandri 5540 (Macul), Santiago de Chile, Chile and ‡Department of Biochemistry and Molecular Biology, Baylor College of Medicine, 1 Baylor Plaza, Houston, TX 77030, USA*

Abstract. Genomes of animals contain between 15 000 (e.g. *Drosophila*) and 50 000 (human, mouse) genes, many of which encode proteins involved in regulatory processes. The availability of sequence data for many of these genes opens up opportunities to study complex genetic and protein interactions that underlie biological regulation. Many examples demonstrate that an understanding of regulatory networks consisting of multiple components is significantly advanced by a detailed knowledge of the spatiotemporal expression pattern of each of the components. Gene expression patterns can readily be determined by RNA *in situ* hybridization. The unique challenge emerging from the knowledge of the sequence of entire genomes is that assignment of biological functions to genes needs to be carried out on an appropriately large scale. In terms of gene expression analysis by RNA *in situ* hybridization, efficient technologies need to be developed that permit determination and representation of expression patterns of thousands of genes within an acceptable time-scale. We set out to determine the spatial expression pattern of several thousand genes encoding putative regulatory proteins. To achieve this goal we have developed high-throughput technologies that allow the determination and visualization of gene expression patterns by RNA *in situ* hybridization on tissue sections at cellular resolution. In particular, we have invented instrumentation for robotic *in situ* hybridization capable of carrying out in a fully automated fashion, all steps required for detecting sites of gene expression in tissue sections. In addition, we have put together hardware and software for automated microscopic scanning of gene expression data that are produced by RNA *in situ*

---

[1]This paper was presented at the symposium by Gregor Eichele, to whom correspondence should be addressed.

hybridization. The potential and limitations of these techniques and our efforts to build a
Web-based database of gene expression patterns are discussed.

*2001 Complexity in biological information processing. Wiley, Chichester (Novartis Foundation
Symposium 239) p 129–149*

Monday June 26, 2000 may become a memorable day in the history of biology.
That day a draft sequence of the human genome was announced by US President
Bill Clinton and UK Prime Minister Tony Blair. Both politicians declared, 'that
there would be an entirely new type of medicine and that disease as we know it
would eventually be a thing of the past'. Such prophecy is likely to exaggerate
the potential of the marked achievement of a human genome draft sequence, but
there is little doubt that knowing the sequence of the entire genome of organisms
opens many opportunities for molecular genetic research and delivers very useful
information into the hands of scientists and physicians. It goes without saying that
the promises made pose a considerable challenge to the scientific community.
Scientists must now deliver tangible results not only to meet the public's
expectations, but also to justify the funds they are being given to pursue genome
research.

    A central theme in genome research concerns the function and interaction of
genes and gene products. Investigations of a number of model organisms have
already unravelled complex genetic regulatory networks. For example,
embryonic development in *Drosophila melanogaster* is now understood in terms of
an orchestrated expression of developmental genes (Lawrence 1992). At the top of
this hierarchy of genes reside the so-called egg polarity genes that coarsely
subdivide the early embryo along its dorsoventral and anterioposterior axes.
Then follows the action of the 'gap' genes that divide the embryo into sub-
regions. Next in the hierarchy are the 'pair-rule' and 'segment polarity' genes that
provide an ever finer segmental subdivision of the embryo. Finally, homeotic
selector genes assign identity to individual segments. As a result of the action of
this hierarchical genetic network, the intricate body plan of an insect is formed.

    These types of findings are largely derived from an approach that starts with a
genetic mutation (e.g. a fly that exhibits abnormal features) followed by molecular
cloning and characterization of the gene that causes the mutation. Hence in this
paradigm the discovery process initiates with a mutation which defines the overt
function of a gene. In the case of genomics, the research strategy is reversed.
Genomics begins with the knowledge of gene sequence and then seeks to define
gene function. Now that we know the sequence of most of the genes, elucidation of
gene function moves to the front. The tools of 'functional genomics' include
characterization of gene expression patterns, structural characterization of gene

products by biophysical methods, investigation of gene product interactions and production of mutants using transgenic methods. Because of its broad scope and significant potential for medicine, functional genomics is moving towards the centre stage of contemporary biomedical research.

This chapter has two parts. Part 1 is a brief discussion of the usefulness of gene expression analysis for understanding gene function. The second part summarizes our efforts in the development of methods to determine and document spatiotemporal gene expression patterns on a genome-wide scale.

Benefits of gene expression pattern analysis

Although each cell of a multicellular organism contains the complete genetic information only a fraction of the genome is expressed in a particular cell or tissue. Gene expression is studied with a broad spectrum of methods including Northern analysis, RNase protection assays, gene arrays, RNA *in situ* hybridization and, at protein level, Western blots and immunohistochemistry. When, where, and how much of a gene is expressed depends on transcriptional and translational regulation, on mRNA and protein transport and on the half-life of gene products. Much effort is put in the development of technologies to determine expression patterns on a genome-wide scale. A widely discussed technology are the microarrays in which DNAs representing many thousand genes are arrayed in spots on an inert surface (Brazma & Vilo 2000, and references therein). Array-based technology for proteins is also emerging (MacBeath & Schreiber 2000). In the case of DNA microarrays, hybridization of labelled cDNA probes derived from tissues of interest enables gene expression profiles of thousands of genes to simultaneously be determined in a single experiment. The disadvantage of this high-throughput technique is a loss of spatial information on gene expression, because the cDNA probes are produced from mRNA of whole tissue extracts.

RNA *in situ* hybridization has been used systematically since the mid 1980s. The principle of this method is simple. Tissues or sections of tissues are probed with tagged synthetic DNA or tagged antisense RNA. Sense mRNA present in the tissue hybridizes with the probe and detection reactions such as autoradiography or immunohistochemistry reveal tissue and cellular localization of transcripts. RNA *in situ* hybridization is carried out on sections or, if the sample is sufficiently small, with whole-mounts.

Antibodies have long been used to localize and characterize proteins in tissue specimens. Because gene function resides in proteins and not in mRNAs, antibody localization of proteins is, in principle, preferable to RNA *in situ* hybridization. Localizations of RNA and protein can also differ, as is the case for secreted proteins. In this instance, RNA is found in cells that express the gene but protein is also present in tissues that may be at a distance away from the site of their synthesis. Examples of this include the peptide hormones that are present in the blood but are synthesized in specialized endocrine glands, and secreted growth factors. In addition to such spatial differences of mRNA and protein distribution, there are also temporal differences. The circadian genes *mPer1* and *mPer2* are expressed in a periodic pattern with a high-point of expression occurring with a rhythm of 24 hours. This cyclical expression is also observed with the corresponding proteins, but the peak expression of protein is delayed relative to mRNA by approximately 6 hours (Field et al 2000). Hence there are cases of differences between spatiotemporal mRNA and protein expression patterns and this must be kept in mind when interpreting mRNA-based expression data. Despite these caveats, it should be emphasized that the presence of a particular transcript in a cell reflects its potential to synthesize the corresponding protein.

There are also notable practical reasons for first focusing on transcripts rather than on proteins. Generating antibodies is time-consuming and costly. Ideally protein with the correct post-translational modifications must be used for immunization and animals are needed for the production of antisera. Among the antibodies presently available only a fraction are suitable for immunohistochemical localization procedures. These technical challenges of antibody work greatly contrast with the ease of using RNA probes, which can be synthesized as soon as a (partial) gene sequence is available using simple and inexpensive *in vitro* reactions.

A vast body of literature demonstrates that *in situ* hybridization and immunolocalization of proteins have been extraordinarily informative in many fields of biology. For example, a good fraction of what is known about the fundamental steps in embryonic development resulted from gene expression pattern analyses in normal and mutant embryos. Gene expression hierarchies that underlie *Drosophila* development were established primarily by studying the expression of gene cascades in wild-type and mutant embryos. In this way it was possible not only to establish gene networks, but also to unravel the cellular basis of these networks. Similar efforts have been undertaken in vertebrate embryos. Here too, RNA *in situ* hybridization has illustrated how cells and tissues interact in the embryo and bring about development. In the nervous system, localization of gene products has uncovered the mechanisms of development and physiology of the CNS. Signals emanating from the dorsal and ventral part of the neural tube specify the identity of neurons in the developing spinal chord. To pinpoint distinct populations of neurons during development and hence to uncover neuronal identity, a series of molecular markers are used that are expressed in a highly restricted, cell type specific pattern in the spinal chord (reviewed in Lee & Jessell 1999). Diagnostic analyses of tumour tissues also makes abundant use of molecular markers.

Development of high-throughput
methods for RNA *in situ* hybridization

Gene expression analysis by histological methods is time-consuming and for the most part is performed manually. This poses a considerable problem when using RNA *in situ* hybridization and immunohistochemistry for genome-wide expression analysis. In fact, without appropriate technology development, this extraordinarily powerful tool is not of much use to functional genomics. To address this issue, we have initiated a technology development program aimed at automation of RNA *in situ* hybridization and automated imaging of expression patterns by light microscopy. In addition, we are currently developing a database that will make gene expression data collected by RNA *in situ* hybridization and immunohistochemistry available to the scientific community through the Internet. Fig. 1 is a flow chart documenting the key steps of the procedure and the text below briefly describes the salient aspects of the individual elements of the flow chart.

Gene selection

Eventually the expression pattern of all genes should be determined because there are presumably very few genes that have no importance. Priority is given, however, to genes encoding proteins that are involved in regulatory and disease processes. This includes receptors, their protein ligands, proteins that mediate signal transduction within cells, channels, transcription factors and their associated regulatory proteins, and extracellular and cell surface proteins that mediate neuronal connectivity, just to name a few examples. It is not uncommon that such molecules can readily be identified based on their primary sequence and characteristic signature motifs and domains. Growth factor receptors contain transmembrane domains, G protein-coupled receptors are characterized by seven transmembrane helices, transcription factors harbour typical DNA binding motifs, extracellular proteins often have characteristic repeat structures such as fibronectin type III repeats, and secreted proteins contain a leader peptide. As the prediction algorithms for protein domains become more robust, selection of particular genes encoding interesting proteins will become more straightforward (Schultz et al 1998).

Probes and tissues

While applicable to any species and tissue type, the focus of our technology efforts are on the nervous system of the mouse. This choice derives primarily from the fact that the brain is the most elusive organ in mammals and it is expected that knowing gene expression patterns at the mRNA and protein levels will help in

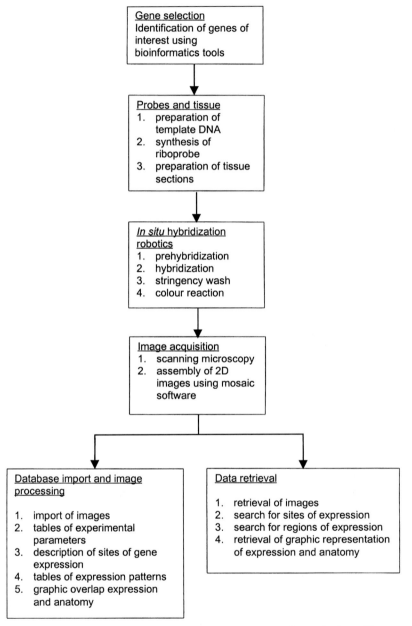

FIG. 1.   Flow diagram of the 'Genepaint' procedure. The term Genepaint is used because the result of the procedure is a visualization of gene expression 'painted' onto the tissue. The procedure is subdivided into several distinct elements each of which can be automated to a considerable extent, including database import and retrieval.

understanding aspects of brain development and function. Although brain functionality undoubtedly depends in an important way on protein interactions within neurons as well as on *trans*-neuronal signalling, knowing the spatial and temporal localization of gene products provides essential base information. The reason for choosing mouse as the model is based on several advantageous features of this organism. First, the mouse is a genetically amenable organism in which genes can be mutated by means of homologous recombination (Joyner 1999). Second, the mouse is suitable to study the development of the CNS and at least simple behaviour can effectively be studied in this organism. There is also marked similarity between mouse and humans at the genetic level. Because of this, the physiology and pathophysiology of normal and genetically affected mice and humans are similar (Habré de Angelis et al 2000). Last but not least, numerous mouse genes and their human homologues have now been identified, expressed sequence tags (ESTs) derived form the mouse CNS exist, and within the next year or so a draft sequence of the entire mouse genome will be available.

Once a gene has been selected for expression analysis, the following steps have to be taken: (1) selection and preparation of DNA template, (2) synthesis of riboprobe and (3) preparation of tissue.

Expressed sequence tags (ESTs) are readily available to generate DNA templates. ESTs represent known as well as presently unknown genes. They are DNA sequences ranging between 500 to 1500 bp and hence may not encode the entire open reading frame of a protein. However, this is not a problem, since a riboprobe corresponding to less than an entire open reading frame is still sufficiently long to confer specific hybridization with the complementary mRNA. Many sequence-verified ESTs are available from commercial or non-profit suppliers and contain T3, SP6 or T7 RNA bacteriophage polymerase binding sites. Because of this feature, EST DNA can readily be PCR-amplified using T3, SP6 or T7 primer sequences which results in a double-stranded DNA template. An alternative to EST-derived DNA templates are PCR products from known genes. In this scenario, RT-PCR is first used to generate a cDNA fragment, which is then cloned into T3, SP6 or T7-containing vectors, linearized and translated *in vitro*. It goes without saying that most if not all of the above procedures can be automated using commercially available equipment.

To synthesize RNA probes used for *in situ* hybridization, DNA templates bounded by RNA polymerase binding sites are used. To detect the probe, a radioactive tag (e.g. [^{35}S]UTP) or a hapten tag (e.g. digoxygenin-labelled UTP) is incorporated into the RNA during an *in vitro* transcription reaction. The advantage of radioactive probes is that they are detectable at low concentration. Their disadvantage is that they are unstable and have to be consumed within a couple of weeks after synthesis. In addition, autoradiography is difficult to automate, uses expensive emulsion, and requires an exposure time of several days or weeks. Hence there is a considerable time-lag between the hybridization step and the instant at which the result of the hybridization experiment can be assessed. These disadvantages have promoted the use of riboprobes that are tagged with a hapten against which high-affinity antibodies have been developed. The most popular epitope — digoxigenin — is detected with an anti-digoxigenin antibody coupled to peroxidase that in turn is detected by a chromogenic assay. The main caveat of this one step amplification is that it is not as sensitive as radioactive probe detection. We found it necessary to add another amplification step based on enzyme-catalysed reporter deposition (Adams 1992, Kerstens et al 1995). Specifically, the peroxidase enzyme attached to the antidigoxygenin antibody is used to activate a biotin–tyramide complex. The activated tyramide reacts with protein molecules in the vicinity of the peroxidase, resulting in a localized cluster of covalently bound biotin molecules. Covalently attached biotin in turn is detected with avidin coupled to alkaline phosphatase, i.e. an enzyme based system that now amplifies the biotin–tyramide signal. It is estimated that this reaction enhances the sensitivity by 100-fold compared to single-step amplification. Also note that riboprobes with digoxygenin tags are stable for months or even years, a marked advantage over radioactive riboprobes.

Because tissue preparation significantly contributes to the sensitivity of a detection method, affects the visual appearance of the specimen and has considerable influence on the signal to noise ratio, several commonly used methods of tissue preparation have been examined. For embryonic and adult mouse sections of frozen tissue have proven to be optimal, but sections of paraffin-embedded tissue can also be used, although tissue quality is not as good as with frozen sections. This is in part due to the fact that tissues are dehydrated in organic solvents, embedded in paraffin and then deparaffinized. These procedures are not only time-consuming but reduce tissue quality. In addition, dehydration reduces tissue size by 20–30%. We prepare frozen sections as follows. Fresh tissue such as mouse brain is collected, placed into a mould filled with cryomount medium and frozen on a slab of dry ice. In this form, tissue can be kept for many weeks without dehydration at −80 °C. Sections are cut in a Leica 3020 cryostat, placed on a microscope slide and fixed in 4% paraformaldehyde. Slides are then stored for weeks or months at −80 °C. Contary to a widely-held belief, cutting frozen sections on a modern cryostat is not more difficult than preparing paraffin sections.

In situ hybridization robotics

The manual execution of *in situ* hybridization, a multiple step procedure, is very time consuming. This problem is exacerbated by the number of samples and genes, which is large in the case of genomics-directed projects. To cope with throughput problems a new technology has been developed. Specifically, slides carrying the tissue sections are integrated into a flow-through chamber (Fig. 2A). The slide is placed inside an aluminium frame with sections facing up. Two 80 μm-thick spacers are positioned along the long edges of the slide, and a 5 mm thick glass plate is placed on top of the spacers. Slide, spacers and glass are clamped together with two metal brackets. A depression milled into the top portion of the glass plate plus the slide form a reservoir (Fig. 2A). This device thus constitutes a small flow-through chamber with a 400 μl reservoir on top, an 80 μm thick chamber housing the tissue (volume: 120 μl) and an exit for solutions at the bottom. Solutions filled into the well enter the narrow hybridization chamber by gravity and remain in place due to capillary forces until displaced by a fresh solution added into the well. Importantly, reagent in- and out-flow is uniform and flow rates are low, reducing the risk of tissue shearing by liquid flow. This flow-through procedure contrasts with manual procedures in which the reagents are pipetted on top of the sections. The sections are then covered with a glass coverslip that has to be removed following completion of the hybridization and immunohistochemical reaction steps. In the case of manual procedures, several of the steps require that slides be moved in and out of solutions, which can result in tissue damage and, furthermore, consumes significant amounts of reagents. Of note, all components of the hybridization chamber are made of heat- and solvent-resistant materials and, with the exception of the spacers, can be reused.

The aluminium frame of the hybridization chamber fits into a temperature-controlled rectangular, protruding platform. Eight such platforms are arranged into a row and six rows constitute a 48-position rack (Fig. 2B). In this set-up, the slide of the hybridization chamber is in direct contact with the surface of the platform permitting efficient thermal transfer. Moreover, platforms and the walls of the racks are hollow, allowing flow-through of liquid delivered by external water circulators. Temperatures typically used in *in situ* hybridization experiments range from room temperature to 55–65 °C. In this range, hybridization chamber temperature is constant with an accuracy of ± 0.2 °C and temperature variation across the rack is ± 0.5 °C. The controller software of the liquid-handling robot can generate temperature profiles by regulating external circulators and valves. Temperature changes in the range of 20–30 °C can be achieved within 5 min or less.

A liquid handling system retrieves and delivers, in a programmable fashion, buffers and reagents from receptacles placed next to the racks. Figure 2C depicts a Tecan Genesis platform whose software has been adapted to the *in situ* hybridization protocol. Using this equipment with four racks (192 hybridization chambers), prehybridization steps, hybridization, stringency washes, chromogenic detection reactions and counterstaining are carried out automatically with little human intervention and supervision. One 'run' takes approximately one day and

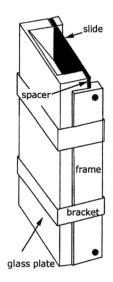

A   B

C

depending on how many sections fit on a slide, a daily throughput of as many as 1000 sections can be achieved.

Image acquisition

The massive amount of data generated by the *in situ* hybridization robot requires effective and automated image data acquisition. The fundamental issue to consider in image data collection is that of resolution. Non-radioactive riboprobes hybridized to mRNA are detected by serial amplification steps (see above) thereby creating a blue-coloured granular precipitate that resides in the cytoplasm. Because the size of an individual grain is in the range of $1\,\mu m$, resolution better than this dimension is not informative. In addition, the subcellular localization of mRNA is usually not informational for studies aiming at definition of sites of gene expression. It should be recalled that the dimension of typical specimens such as a mouse brain is in the range of millimetres or centimetres. Thus one needs to use an imaging technique that spans the three to four orders of magnitude between the dimension of the signal ($1–10\,\mu m$) and the dimension of the specimen (~ 1 cm).

One approach is provided by optical scanning devices, as has recently been illustrated by Hanzel et al (1999). These investigators used a confocal digital microscanner for high-throughput analysis of fluorescent specimens. This strategy allows a resolution of approximately $5\,\mu m$ per pixel and scanning of a large field (entire sections) within a time frame of minutes.

An alternative strategy of data acquisition is based on a compound microscope equipped with a scanning stage that stepwise and accurately translocates the specimen in front of the objective. At each step, the specimen is focused and an image is acquired with a CCD camera. The reproducibility of stage movement is $\sim 0.75\,\mu m$ and the resolution of the x–y translation is $\sim 0.025\,\mu m$. A segmental data collection procedure requires software that subsequently assembles individual images into composites. The advantage of a microscope-scanning

---

FIG. 2. *In situ* hybridization robot. (A) A hybridization chamber consisting of a standard microscope glass slide (depicted in black) loaded with sections, two spacers, and a glass plate that are assembled onto an aluminium frame and held together with two spring clamps. (B) Hybridization chambers are placed into a temperature-regulated rack. The photograph shows the first row of the rack occupied by hybridization chambers (arrow). (C) A TECAN platform housing two such racks plus various containers of different volumes in which the solutions used for hybridization, washing and colour reaction are stored. Note the eight dispensing pipettes in the back of the platform (arrow). These pipettes take up and then deliver solutions into the reservoir on top of each hybridization chamber (see A). There is sufficient space on the Genesis 200 platform to accommodate four racks plus solvent containers.

stage-based system is that it generates high-resolution data of the expression signal and of the underlying tissue histology. Using a 440 000 pixel CCD camera, a ×5 objective and a ×10 eyepiece provides a resolution of approximately 3.5 μm per pixel. This resolution can readily be increased to below 1 μm per pixel when a ×20 lens is used (e.g. Fig. 4B, D).

The equipment used in our laboratory consists of a Leica microscope, a motorized Märzhäuser stage that accommodates up to eight slides, a Leica electronic focusing motor, a JVC CCD camera and a PC controller using the image analysis software QWin V2.3 from Leica (Fig. 3A).

Data acquisition is initialized either by a prescan during which the location of each section on the slides is empirically determined or by scanning within a predefined area. In the prescan mode the stage meanders stepwise across the array of slides and at each step an image is captured. An integrated intensity of each of these images is calculated and compared against a preset threshold value. Only those images in which the integrated intensity is greater than the threshold value are considered to represent tissue. Rectangular frames, termed regions of interest, are calculated, each of which encloses one section. The prescan thus results in the definition of the position of each of the sections on a slide. The subsequent main scan collects images only in the regions of interest. The main scan is more time-consuming than the prescan, since the acquisition of high quality images requires determination of the optimal focal plain. This is achieved by an autofocus routine executed prior to image capturing.

In the template scanning mode all tissue sections are collected into rectangular windows defined by a mask that is transiently fastened to the slide (Fig. 2B). Using this device allows sections to be placed at predefined positions. Hence images are captured only within this predefined rectangle. As is the case for the prescan mode, an autofocus routine is executed prior to image capturing.

The prescan/mainscan procedure and the template procedure each have specific advantages and drawbacks. The first strategy does not require that sections be placed to a particular position, because the region of interest is determined empirically. The prescan takes time, making this mode about 50% slower than the template mode. In addition, if sections are weakly stained, setting a detection threshold can be difficult. If the threshold is set too low, the procedure fails to distinguish between the section and background and if the threshold is set too high, sections fail to be recognized at all. The template mode does not have this complication, but requires some skill in placing sections at defined positions and depends on having appropriate masks.

In either mode of data collection, individual images are stored as bitmap files. They are named using a so-called speaking key 'ImageName @ Xposition @ Yposition' (see Fig. 3C). The information on images is stored in an ASCII database that also includes data on specimen characteristics such as the identity of

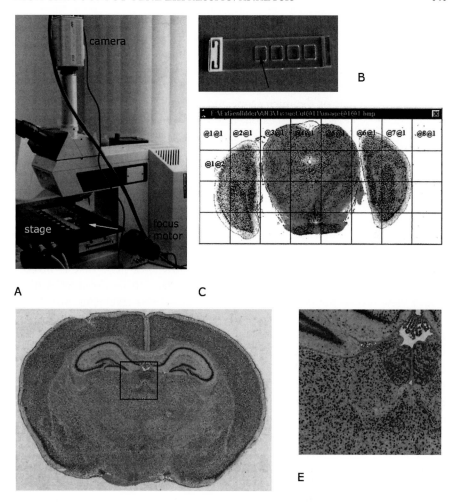

FIG. 3. Image data collection system. (A) Leica microscope equipped with an analogue camera, an eight-position scanning stage and an autofocus motor. (B) Acrylic template used for proper positioning of sections on slides. One of the four windows of this mask is indicated by an arrow. (C) Image nomenclature used by the 'Exgen' software package. Each square corresponds to one captured image. (D, E) Composite image assembled using mosaic software. The frame represents one image that is shown at higher magnification in (E).

the gene whose expression is being determined, name of the operator, thickness of and distance between section, etc. The platform-independent database, an open data format and the 'speaking key' annotation of the images make import of the data into another database fairly simple.

As illustrated in Fig. 3D, a section is composed of several images. An image set may consist of as few as four elements and be as large as 50 or 100 elements. The image set (mosaic) is created using mosaic software that assembles individual images to a composite. To create a composite of adjacent but not overlapping images requires accurate stage movement and a precise alignment of the camera with the optical axis of the microscope and the x and y axes of the stage. This alignment is achieved using custom-made software and mechanical centring devices. Fig. 3D shows a composite image and a black square outlines one single element of ~ 0.8 Mb, shown at higher magnification in Fig. 3E. To capture one image element takes ~ 3 seconds, i.e. the composite shown in Fig. 3D was collected in a few minutes. Composite images can be stored in a variety of formats including JPEG, TIFF and as Bitmaps. TIFF files for a single section of a mouse brain typically amount to 25–50 Mb.

Sample data generated by the system described above are shown in Fig. 4. Figure 4A depicts the pattern of expression of the transcript encoding the γ subunit of *calmodulin–calcium-dependent protein kinase II* (*CaMKII*), a gene that is broadly expressed in the mouse brain. The specimen is a 20 μm frozen section. A high-power view reveals the details of expression at a cellular level (Fig. 4B). It can readily be seen that expression data have single-cell resolution, a distinguishing feature that cannot be achieved with radioactive riboprobes unless cells are very large (e.g. Purkinje neurons of the cerebellum). Figure 4C depicts the pattern of expression of the circadian clock gene *mper2* on 7 μm thick paraffin sections, a gene that is almost exclusively expressed in the suprachiasmatic nucleus. Even here, it is clear that single neurons are seen, although the overall quality of the tissue is somewhat lower than seen with frozen sections. The data shown in Fig. 4D depict the expression of *Nurr1* (a nuclear hormone orphan receptor; Xiao et al 1996, Castillo et al 1997) in a 20 μm frozen section through the cortex. This illustration depicts *Nurr1*-positive and negative cells and demonstrates that detection of expression based on digoxygenin-tagged riboprobes can readily distinguish between expressing and non-expressing neurons. All data shown in Fig. 4 have low background staining, yet background is sufficiently high to visualize cells that lack transcripts eliminating the need of having to counterstain sections. In summary, non-radioactive probes are excellent tools to detect expressed genes in (frozen and paraffin) sections and are in many respects superior to radioactive probes.

Database

Numerous databases containing biological data are accessible through the Internet (e.g. Discala et al 1999, Ringwald et al 2000). Web-based databases are an excellent means to efficiently retrieve and hence disseminate scientific data. Automated

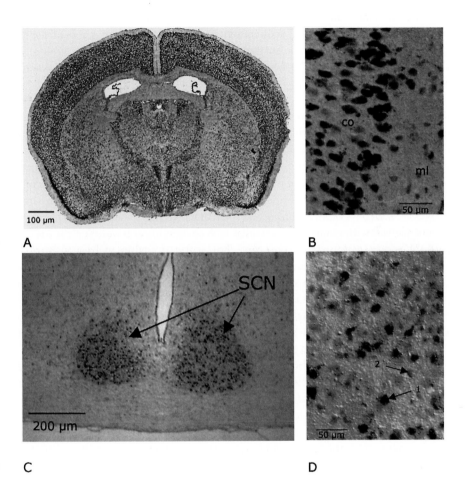

FIG. 4. Examples of images of expression patterns created with the Genepaint system. (A) The *calmodulin–calcium-dependent protein kinase II* gene (*CaMKII*) is broadly expressed in this coronal frozen section through an adult mouse brain. This image is a mosaic consisting of ~40 individual images. The section is 20 μm thick and the image was taken with a ×5 objective and a ×10 eyepiece. (B) High-power view of outer layers of the cortex (co), showing numerous *CamKII* expressing neurons. ml, layer 1 of the cortex. The section is 20 μm thick and the image was taken with a ×40 objective and a ×10 eyepiece. (C) Photograph of a 7 μm paraffin section hybridized with *mper2* riboprobe. Strong expression is seen in the suprachiasmatic nucleus (SCN). Note individual neurons in this preparation. The image was taken with a ×10 objective and a ×10 eyepiece. (D) This 20 μm frozen section was hybridized with a *Nurr1* antisense riboprobe and reveals expression of this gene in a subset of cortical neurons. Arrow 1 points at a positive cell whereas arrow 2 indicates a non-expression cell. The image was taken with a ×40 objective and a ×10 eyepiece.

image acquisition by the procedures described in this chapter results in a large quantity of data from a variety of specimens and hence a database is also a powerful tool to track data acquisition and for data storage. A database is currently being developed that can provide these functions. As outlined in Fig. 1, the task of the database can be divided into two segments: data import/processing and data retrieval.

Data acquisition itself is a multistep process and one critical function of the database is to provide information on the specimens used (species, strain, fixation methods, histological methods), the sequences of the genes analysed (e.g. GenBank accession number, cloning vector, sequence of template) and experimental parameters of hybridization and of detection. A second critical task of the database is the storage of images. As pointed out above, the average size of a brain section stored in TIFF format is approximately 30 Mb. We plan to store such images on a Web server in a compressed form in which they can be readily retrieved by and viewed with commercially available Web imaging software such as MGI's ZOOM Server (*http://www.mgisoft.com*). This software enables website visitors to zoom into images and thus reveal details commensurate with the high resolution of the primary image data provided by scanning light microscopy. An additional feature of the database is a summary description of the sites of gene expression in tabulated form. Image analysis procedures will be developed that are capable of an automated identification of sites of expression and able to relate these sites of expression to standard brain anatomy maps.

A detailed map of mouse brain anatomy currently exists only for the adult mouse brain (Franklin & Paxinos 1997). Until now, a limited amount of information has been compiled into atlases of embryonic or postnatal murine CNS, but efforts are being undertaken to close this gap (Valverde 1998, Jacobowitz & Abbott 1998, Kaufman et al 1998, Brune et al 1999). An important facet of this effort is the development of an appropriate and systematic nomenclature of cells and tissues (Altman & Bayer 1995, Bard et al 1998). This anatomical nomenclature should be hierarchical so as to facilitate a search for gene expression patterns in structures and substructures. Another requirement of the nomenclature that it needs to reflect the developmental history of a tissue. The CNS, for example, arises from a simple tube-like structure which becomes progressively more complicated and eventually ends up as a structure consisting of well over 1000 distinct anatomical structures. To have a nomenclature reflecting the developmental hierarchy of tissues would be extremely beneficial.

A key task the database has to accommodate is the retrieval of the data using parametric searches. For example, it is planned that the database can be queried for the sites of expression of specific genes and for regions of overlapping expression patterns of two or more genes. Pilot tests have been undertaken to retrieve images of gene expression using MGI's Zoom server software. Retrieval of images of resolution and quality equal to the primary data in the time-frame of seconds was achieved worldwide using standard Web browsers such as Netscape or Explorer without the need to instal any additional software. By downloading an appropriate freeware plug-in, it is possible to move around within images and zoom into images reminiscent of the way this is done in the microscope.

In summary, we have developed procedures that allow an efficient production and analysis of gene expression patterns by *in situ* hybridization on tissue sections. This instrumentation will make it possible to analyse large numbers of genes in a reasonable amount of time and thus will allow us to create spatial maps of gene expression in mouse brain. These maps will not only be established for the adult brain, but also for embryonic and early postnatal CNS. This endeavour in combination with other technology is likely to shed light on the genetic and biochemical signalling networks that underlie brain development and function.

Acknowledgements

We thank Claus Ebert, Sabine Bergmann and Heike Krause for expert technical assistance. This work was supported by funds from the Merck Genome Research Foundation and The Max-Planck-Society for the Promotion of Science.

References

Adams JC 1992 Biotin amplification of biotin and horseradish peroxidase signals in histochemical stains. J Histochem Cytochem 40:1457–1463

Altman J, Bayer SA 1995 Atlas of Prenatal Rat Brain Development. CRC Press, Boca Raton, FL

Bard JBL, Kaufman MH, Dubreuil C et al 1998 An internet-accessible database of mouse developmental anatomy based on a systematic nomenclature. Mech Dev 74:111–120

Brazma A, Vilo J 2000 Gene expression data analysis. FEBS Lett 480:17–24

Brune RM, Bard JB, Dubreuil C et al 1999 A three-dimensional model of the mouse at embryonic day 9. Dev Biol 216:457–468

Castillo SO, Xiao Q, Lyu MS, Kozak CA, Nikodem VM 1997 Organization, sequence, chromosomal localization, and promoter identification of the mouse orphan nuclear receptor Nurr1 gene. Genomics 41:250–257

Discala C, Ninnin M, Achard F, Barillot E, Vaysseix G 1999 DBcat: a catalog of biological databases. Nucleic Acids Res 27:10–11

Field MD, Maywood ES, O'Brien JA, Weaver DR, Reppert SM, Hastings MH 2000 Analysis of clock proteins in mouse SCN demonstrates phylogenetic divergence of the circadian clockwork and resetting mechanisms. Neuron 25:437–447

Franklin KBJ, Paxinos GT 1997 The Mouse Brain in Stereotaxic Coordinates. Academic Press, New York

Hanzel DK, Trojanowski JQ, Johnston RF, Loring JF 1999 High-throughput quantitative histological analysis of Alzheimer's disease pathology using a confocal digital microscanner. Nat Biotechnol 17:53–57

Hrabé de Angelis MH, Flaswinkel H, Fuchs H et al 2000 Genome-wide, large-scale production of mutant mice by ENU mutagenesis. Nat Genet 24:444–447

Jacobowitz DM, Abbott LC 1998 Chemoarchitectonic Atlas of the Developing Mouse Brain. CRC Press, Boca Raton, FL

Joyner AL 1999 Gene Targeting: A Practical Approach, 2nd edn. Oxford University Press, Oxford

Kaufman MH, Brune RM, Davidson DR, Baldock RA 1998 Computer-generated three-dimensional reconstructions of serially sectioned mouse embryos. J Anat 193:323–336

Kerstens HMJ, Poddighe PJ, Hanselaar AG 1995 A novel *in situ* hybridization signal amplification method based on the deposition of biotinylated tyramine. J Histochem Cytochem 43:347–352

Lawrence PA 1992 The Making of a Fly: the Genetics of Animal Design. Blackwell Science, Oxford

Lee KJ, Jessell TM 1999 The specification of dorsal cell fates in the vertebrate central nervous system. Annu Rev Neurosci 22:261–294

MacBeath G, Schreiber SL 2000 Printing proteins as microarrays for high-throughput function determination. Science 289:1760–1763

Ringwald M, Eppig JT, Richardson JE 2000 GXD: integrated access to gene expression data for the laboratory mouse. Trends Genet 16:188–190

Schultz J, Milpetz F, Bork P, Ponting CP 1998 SMART, a simple modular architecture research tool: identification of signalling domains. Proc Natl Acad Sci USA 95:5857–5864

Valverde F 1998 Golgi Atlas of the Postnatal Mouse Brain, Springer Verlag, Heidelberg

Xiao Q, Castillo SO Nikodem VM 1996 Distribution of messenger RNAs for the orphan nuclear receptors Nurr1 and Nur77 (NGFI-B) in adult rat brain using in situ hybridization. Neuroscience 75:221–230

DISCUSSION

Sejnowski: How many genes are you going to be able to do per week?

Eichele: In the mouse brain expression analysis, we intend to look at four stages, including adult. We will do selective sections; perhaps 15 for the adult brain and take a corresponding number for the developing stages. This will give us a set of about 20 slides for one gene. If we have 200 positions, this will enable us to do 10 genes per day, which works out at 50 genes a week. I suspect we will have to do duplicate experiments, so the theoretical output of 2500 genes per year will probably not be reached. Like sequencing, it is just a matter of money, because the technology is basically there. We plan to use ESTs as templates for making our RNA.

Sejnowski: What is the behavioural status of the mice before they are sacrificed?

Eichele: We try to use as uniform conditions as possible. We use the C57BL6 strain of mice. For the adults, we will take them at 8 weeks of age and keep them under constant conditions, sacrificing them at a specified hour of the day. Of course circadian and sex-specific genes will sometimes be missed.

Brenner: There is a mini-problem here, which should be tackled, in the retina, where you can see how deep you need to plumb specificity. For example, it is claimed that there are 22 kinds of amacrine cells. Are these all different cell types in the sense that they represent disjoint sets of gene expression combinations? One distinct amacrine cell is called the starburst amacrine cell. Some starburst amacrine cells have been shown to have acetylcholine. The question is, are all acetylcholine cells starbursts, and are all starburst cells acetylcholine cells? We know that rods and cones are different, and that there are three kinds of cones based upon the visual pigments that they produce. The retina will allow us to do things even at the biochemical level much faster than your high-throughput, brute force approach, and provides us with a simpler way of getting real information. There must be simpler ways of doing this and getting real information.

Eichele: You are asking two questions here. I agree with what you said about the retina, and similar thoughts apply to the spinal cord. None the less, we somehow need global expression information for the extensive CNS. Recall, the issue of cell fate is just one problem: gene expression has additional important facets. Yes, our efforts are a 'stamp-collecting' approach, just as sequencing genes is. But a systematic approach does provide a base of information.

Brenner: If you can get all the stamps it would be terrific. The trouble is that it is going to be very difficult. There will be rare genes expressed in rare cell types that are going to be critical, and which you may not be able to find this way. If it does not aim at completion, it isn't going to be worthwhile. One has to think very carefully about the uniqueness of a cell type in the brain, and I think the retina is a well defined way of starting. Let me mention another approach. Brains could be dissociated into cells, and an antibody could be used to collect all the cells that express the muscarine type I receptor, for example. A total analysis of everything expressed in these cells could then be done and correlations established. More antibodies could be used to select more complex subsets of cells. The same antibodies could then be used to localize the cells in the brain.

Sejnowksi: Dick Masland has done this in the retina (MacNeil & Masland 1998). He has gone through systematically and used all the known neurotransmitter probes. He has made a lot of progress on the amacrine cells. He is finding many more cell types than people had discovered previously, but it is very hard going.

Eichele: Alternative approaches to the one I proposed are certainly worth pursuing. I am just saying that systematic strategies are powerful and make no assumptions about outcomes. Some people may not look at the brain as an issue just of cell fate: there are other key issues such as cell migration and neural degeneration. The other point I wanted to respond to is an issue raised by Dr Brenner, concerning rare meassages. Certainly, for the adult brain I can envision a situation where rare messages are very important. It turns out that in developing embryos, many key signalling molecules — such as transforming growth factor β, bone morphogenetic proteins and Sonic hedgehog — are always expressed at reasonably high levels in one or more cells. Although these regulatory molecules are on the whole (i.e. if you average over the entire organ) expressed at low levels, they are usually expressed locally at high levels and thus are easy to detect. So I am not too worried about the detection problem on the basis of my experience in embryology.

Berridge: In *Caenorhabditis elegans*, where we know all the neurons, how many neurons are similar? How much divergence is there?

Brenner: You can unify them by transmitters into subsets. They have functional connotations. But *C. elegans* has very few cells, so quite often there are unique single cells.

Berridge: I am surprised by this figure of 500 cell types in the brain.

Eichele: This is a guess.

Brenner: How many cell types do you think there are in the body?

Eichele: I would estimate 1000–2000.

Brenner: It is about 200, if you look at a classical textbook of histology. I think the brain must have at least the same number.

Sejnowski: Francis Crick went through this exercise a few years ago, and came up with about 1000.

Berridge: How different are these? Are there groups like pyramidal cells with subtle variation, which are all classed as different?

Brenner: If you use the technology we have to look at one cell type, such as a monocyte changing into a macrophage, and you look at how many of the genes change by a factor of two or more, it is something of the order of 30 000 genes. Some of these can be made sense of and are trivial because this is an exponential cell which has to come out of the growth phase, but it is an enormous number of genes changing in this one process. You can see the big things that turn out to be 1% of the entire message. The analysis of this is quite difficult.

Berridge: Presumably, this would not show up in this sort of screen. Gregor Eichele is doing a screen that is quantitative in nature.

Eichele: It is semi-quantitative. It is also questionable whether mRNA levels actually reflect protein levels. I could have told you of several examples where that is not the case.

Brenner: I think the mRNA is important and I think this argument about whether mRNAs reflect protein levels is neither here nor there. It reflects the protein-coding capacity of the cell which you want to know: can this cell make this channel or not?

Berridge: From a physiological point of view, in the cortex we have a series of columns, which helps to simplify matters. What I would like to know is if you take something like the cerebellum, there appear to be relatively few cell types that have been recognized for making up the circuitry. The same applies for the hippocampus. Now if we come to the cortex, are you telling me that there need to be 500 individual cell types to make up the circuitry?

Sejnowksi: I need to explain a bit about how this is done. It depends on topography. For example, it is known that there are cells that tile the retina, in the sense that their dendrites don't overlap very much, but they cover the entire area. If this is defined as one cell type, all cell types have similar tilings.

Brenner: It is about 50 in the retina based on the minimum member.

Sejnowski: There are problems in the cerebellum: it is not true that all Purkinje cells are the same. It is known that there are zones where different peptides are expressed in different subsets of cells.

Eichele: If you look at the expression patterns of genes such as *zebra*, you see stripes of expression in the cerebellum. Of course, this doesn't mean that these stripes represent different cell types, but they might.

Sejnowski: These are stripes that are very regular. It is known that these stripes project to different locations within the inferior olive and so forth. You have to look at more than just expression in order to define a function. There is something that is even more disturbing to me: you brought up circadian rhythms as a clear case where the gene expression is going to change in a regular way. It turns out that any disturbance to the animal causes the intermediate-early genes to be activated. This means that small changes in the previous history of the animal's experience are going to be causes of activation in whole sets of different genes. With long-term memory, there are presumably changes in gene expression that are going to last for even longer, that have to do with the exact details of how the animal was handled in the past. Rusty Gage at the Salk Institute in La Jolla recently discovered that there are new cells being born in the dentate gyrus of the hippocampus every day, and their survival depends on the experience of the animal. If the animal runs on a treadmill more of those cells survive than if it doesn't. If that is the case, there must be lots of changes occurring in the patterns of the gene expression depending on exercise.

Brenner: A lot of our thinking is based on the haemopoeitic system, in which there are different kinds of granulocytes and lymphocytes. Are you going to call a cell which has been subjected to stress and has a new gene expression pattern a different cell type? This is why I am so keen on classifying them according to their receptors. The receptor tells me something about functionality. And a lot of the antibodies are available.

Eichele: One of the applications of the kind of approach we are using in the lab is for screening genes. For example, we have done a lot of two-hybrid screens with circadian proteins. We have a specific hypothesis and then we use *in situ* hybridization for co-localization of prey and bait. This approach can be implemented in any lab very efficiently to sift through screens in a week or two, which without robotics would previously have taken a year.

Reference

MacNeil MA, Masland RH 1998 Extreme diversity among amacrine cells: implications for function. Neuron 20:971–982

General discussion II

Understanding complex systems: top–down, bottom–up or middle–out?

Sejnowski: In this general discussion I'd like us to address the issue of the best approach for trying to understand complex systems. Do you start at the top or the bottom, or do you jump in somewhere in the middle?

Noble: Earlier on Sydney Brenner made a comment about preferring to go middle–out, rather than top–down or bottom–up. It helps to ask the question, what is wrong with the bottom–up and top–down approaches? The bottom elements of a biological model will be components such as proteins. We know that there are large numbers of these. We may then want to model a subcellular system, such as the Ca^{2+} oscillator, so we bring together a group of proteins and make a model of these. In my case the next level might be modelling different cells. You can think of it in your own field in a different way. Then to build up to the next level, you will want to bring together a lot of these cellular elements. As we go on up, more of these lower level components start to come in. One might think that it is obvious how to proceed from the bottom to conflate at each level, by bringing together a group of processes and representing them by simpler equations and simpler ways of thinking about the next step up. This, I suggest, is a reasonable caricature of the bottom–up approach. The top–down approach takes the view that we look at some piece of functionality at the top, such as the heart beating or the pancreas secreting, and we try to work the other way and break things down into components at each level, hoping that somehow or other it will connect with the bottom–up approach. I am going to argue that, first of all, there is no guarantee that they will connect, and moreover both approaches are impossible. Why is this the case? There is no one way in which you partition your dividing lines amongst all the components. You choose what you bring together as a set of components in one model; but someone else measuring another process might want to take in another group, also. This means that we end up with a different element at the next level. There are many ways in which you could break up such a multilevel structure. This is even more true given the fact that we now know that genes play roles in many different functionalities. We can't even think therefore that we have a unique way of going up. I would argue that it is impossible to go completely bottom–up because there is no unique way of doing it. The obvious difficulty with the top–down approach is that we will not know what is the best way of breaking the upper levels into components at a lower level, so that it would meet any one of the bottom–up approaches. I am going to argue that, whether he meant it or not, Sydney's comment that he preferred middle–out is actually totally necessary. There is no alternative.

Brenner: From my point of view, you are already starting at the middle: you are starting with proteins. That's fine; I like that. I'd go a little bit higher, actually.

Noble: I would argue that even just starting with proteins is not going to work.

Brenner: That's the middle. Starting with genomics is what I call the bottom–up approach. I think this is impossible.

Noble: The comment about multiple ways of dividing elements up takes me to the next remark I wanted to make. It seems to me that whichever of these intermediate levels you might be concentrating on, you are going to need physiological insight: you need to know what function you are reproducing. What I'm saying is, I hope, absolutely obvious. There is a very important conclusion that comes from this, which is that not only will it lead to you dividing the world up in different ways at this level according to the functionality you seek to reproduce, this will also happen all the way up and down. Therefore, there will inevitably be many models of the heart and pancreas, for example, and it all depends on what you want to do with them. This leads on to another question: how can the different models be mapped onto each other? We have been thinking about this a lot in relation to simplifying complex models.

Fields: Isn't there another way to approach this problem? This is to view the system as having an input and an output. In this, we see the complex system as a sort of black box, and we look at the input function (which may be action potential) and the output function (which may be gene expression). When we understand the correlation between the two, we can go in and measure a unit or system within that box. We can investigate what happens when we perturb this component of the overall system, and then go in and measure the input and output function again to see how it has changed. We can continue this process until we can reconstruct the complex system.

Brenner: That is pretty a fair approach, but you have to partition the problem. Some of us are still very interested in how the representation of an organism in the genome changes in the course of evolution. We think this is very important because we know organisms succeed by trying out genome variations to see whether they work. I believe that most biologists are committed to this rather unique property of biological systems in the world of natural complexity: the existence of an internal description of the organism written in the genes. In the 1960s, when we thought that all the problems of molecular biology were solved, some of us decided to go to more complex biological systems. Seymour Benzer decided he would try to connect genes with behaviour. His aim was to use mutations of behaviour to 'parse' the genes of the organism. I, on the other hand, thought that this transformation would be too complicated to understand, and that there certainly would not be a mathematical function to transform gene space into behavioural space. I felt very strongly that there were really two separate problems. First of all, we need to build the object that 'behaves': this is where the genes fit in, and this is one problem we can try to solve. Then the second problem of how such a gene-constructed machine generates behaviour can be studied independently. So, all of this grand genetics collapses into two classical problems—one of development and one of physiology—but it turns out they are united by one thing: structure. Thus the key to understanding the nervous system is anatomy. You want to know what this is and how it is generated by development, and then you want to know how it works. Now that we have genome sequences we can go back to address the old questions with the componentry defined. For my part, the middle area between the organism and the genes consists of cells. What I want to know is how the genes get hold of cells. This discussion of cell types is terribly important, because it will tell us about the true complexity of that space. This is my 'middle', and the cell is my unit, because I can look outwards from the cell to physiology and inwards to molecules.

Laughlin: I would like to return to the more mundane problem of the black box. Douglas Fields is exactly right: we are not going to get anywhere unless we can associate in a deterministic way a particular input with a particular output, because that is telling us what the box is doing. The trick is to pick the right set of inputs. We have heard that some of these boxes are non-linear and they are adaptive, and we have heard before that we must look at a cell signalling system within the context of the correct set of inputs. This is where we need to go back and look at the physiology of the system as a whole. You can't pick your inputs out of thin air.

Sejnowski: There is a problem with this approach: it assumes that what comes out is a consequence of what goes in. In other words, there are autonomous events occurring within brains that produce behaviour and create information. Therefore, if your whole analysis is based upon this input–output, stimulus–response behaviourist conceptual framework, you are doomed. You are missing out on something really important about how cells work and how brains work: they are not stimulus–response devices.

Laughlin: That seems to imply that there are things that happen without any preceding events. I can't believe that.

Sejnowski: They are all oscillations: intrinsic mechanisms such as circadian rhythms. What do you think happens at night when you fall asleep? Do you think that what is happening in your brain is a consequence of what is going in at that moment?

Laughlin: Your oscillator happens to be entrained. We shouldn't argue too much about this, but I suppose that you are saying that there are closed systems where the output is also the input. We can deal with those. They are not going to be very complex systems, because they are not connected to anything else. The really complicated systems are open.

Sejnowski: No, I am making a different point. Yes, we do have sensory surfaces that take in information, and we do respond to the world. But in addition to these, we are also generating internally very complex signals that we are imposing on the world.

Berridge: That is part of the black box.

Sejnowski: What I am arguing against is the traditional stimulus–response analysis: that the way that you analyse the black box is by giving it a stimulus and looking at the response. If we do this, we'll miss part of what is in the black box.

Laughlin: I am saying that you have a knowledge of the processes, which tells you that given something happening in the system, this will be the consequence. It is a good local description, which I think is what the idea of the black box was aimed towards.

Segel: I agree. There will inevitably be many models of the heart, for example. I wrote something about this sort of thing a few years ago (Segel 1995). You will ask different questions and you will need different models for the different questions. Some of the questions will require detailed submolecular models at a high level. Some will need input–output models, and others not. If you want to understand an automobile, you may be talking about the engine and wheels and transmission at one level to answer certain questions. At another level, probably the biggest cause of failure of automobiles where I come from in New England is rust, which is due to a fault in the paint, which involves a different set of questions and a corresponding different level of modelling. The more complex a system, the more models will be needed.

Prank: Choosing the right input in the black box approach is important. If we put in the wrong input we won't answer the question we have addressed. One important example is a study on the precision of spiking in neocortical neurons. It has been demonstrated that the same constant depolarization current leads to imprecise spiking if injected in a repetitive way to the same neuron, in contrast to a fluctuating current consisting of Gaussian white noise which had the same mean amplitude as the constant current (Mainen & Sejnowski 1995). The white noise current input produced a highly reproducible precise temporal pattern of spiking.

Dolmetsch: In theory, it is possible to have some sort of unified model. After all, cells do work, and in principle there is a single way of describing everything. What you have described is the pragmatic approach: in practice we can't model what every protein is doing, but in principle we could. In principle it should be possible to model everything from the bottom. In fact, it should form an assembly.

Sejnowski: Let's give you your dream. Suppose someone actually went out and spent the entire budget of the world to create a perfect model of a cell, down to the last molecule. You turn it on, and lo and behold it works just like a cell. Every single experiment you do with the model is just the same as in a real cell. What have you learned? If your model is just as complicated as the cell itself, I would claim that you have learned nothing.

Dolmetsch: I disagree with you. The whole point behind having a model is to be able to change parameters that you cannot change experimentally. If you cannot do that, then you don't need the model.

Brenner: You have to distinguish between an imitation and a simulation. The difference is that the simulation is couched in the machine language of the object being simulated. You could write a model that might explain the heart but it uses radar waves that are reflected off surfaces. This will imitate the behaviour of the heart, but it is not a simulation, because it is not written in the machine language of all that stuff that is in the heart. That will involve computation, because what you are asking is how the components generate the next level.

Sejnowski: Are you arguing for or against the bottom–up approach here?

Brenner: Middle–out. The bottom–up has very grave difficulties to go all the way. Denis Noble's bottom is my middle.

Sejnowski: The arguments you have both made are very cogent ones that question the very enterprise itself; whether it is even possible. But what I am saying is, let's go beyond that. Suppose that someone took all the effort and discovered every molecular component of the cell. I really don't think that this, by itself, will be very informative. Perhaps you can do some experiments with a model that couldn't be done with a real cell. With the techniques improving all the time, I think it will be a lot cheaper to do the experiments on real cells.

Dolmetsch: What is the goal? Why do you want to have a model? One reason might be that it allows you to test things more rapidly or cheaply once you have the model. For example, why would pharmaceutical companies be willing to pay lots of money for Denis Noble's model of the heart?

Sejnowski: I love models; I work with them every day. It is the particular type of model that you are advocating which I am questioning. In physics, the 'model' isn't an exact replica down to the last detail. In fact, it is not a model unless you throw something away. It means that you have picked out a subset of all the things there, and you have been able to replicate some behaviour of the system. It is proof that you have, in a sense, identified the critical variables or relationships responsible for the complexity that you are observing. This is when you learn something: not when you have everything there, but instead just a few things and you still get the behaviour.

Dolmetsch: What you really want is for a model to be able to predict an experimental result. If it is going to be predictive, it doesn't matter whether you have all the details.

Sejnowski: A model does more than just make predictions; it means explaining things and understanding, as well as predicting.

Noble: I would go along with you in saying that models have to be partial representations of reality, if you are going to achieve some understanding. This partly comes from the fact that even if we did build a 'supermodel', we would have difficulty understanding it. Then we would have to make partial models of this supermodel in order to understand it. I also suspect that it is possible to prove that what Ricardo Dolmetsch is proposing — delightful though it might be — is impossible. This is a matter of computability. Suppose we say that we want to represent every molecule in the system and compute it. What are we going to use for this? A molecular computer (see Reed & Tour 2000)? Is it possible that the world has got enough stuff in it to compute the world?

Dolmetsch: This was just a thought experiment. It is probable that every single molecular interaction is not computable. But what I meant is if we could start at a fine level, such as the level of proteins. It is conceivable that you will get enough detail at the level of proteins that it will somehow assemble into the right thing.

Sejnowksi: There is another issue involved here, and this came out of Denis Noble's paper — the issue of scale. The type of model that is appropriate for one scale is quite different from the type of model that will be used at another scale. If you are dealing with single proteins for the channels, you might use a Hodgkin–Huxley model. But it is not clear what mathematics you want to use if you are describing organelles or molecular machines. In the case of neurons, for example, when I model entire neurons with a compartmental model, I am dealing with finite elements: I am looking at spatial relationships and the coupling between them. Given that we have many different scales, and many different types of models, how do we join those models together? In other words, what is relevant at one level that we want to include at the next level as a parameter? This will be an important issue, and engineers know this already because they have to deal with simulating a Boeing 777 and they can't just use one model. They struggle with this issue of scale, and finding the appropriate way of matching the details at one scale with the variables at the next one.

Noble: You have posed two distinct questions. One issue is how we map very different types of model, and the other is with regard to making models talk to each other and how we avoid the tower of Babel in the world as we try to link things together. It may seem obvious, but in the end, we need to get away from having to write computer code. We want to be able to deal just with the mathematics. Somehow or another, software has to be developed that will enable us all to code in a way that is invisible: so what we can talk about is the equations we use, whether we code in one language or another, or use one type of software development or another. It seems to me that internationally we have to coordinate to ensure that there is this kind of development.

Brenner: In other words, you need a workbench for this kind of work. It is not a database, which seems to be all that people want.

Noble: We need a platform technology. As a minimum requirement this would have to include equation editors and something like XML to act as a platform (see for example *www.physiome.com*).

Sejnowski: These are issues that have been explored in great depth in computer science: what you need is a software layer within a hierarchy. For example, the Windows operating system has a user interface that is quite different from what is going on underneath when manipulating your files. What you are describing is the need for the equivalent of the Windows user interface that will free us from having to wiggle all the bits down at the bottom layer. However, for something like this to happen we would need a Microsoft equivalent.

Brenner: You need a commitment, and I don't see this going into every home in the world!

Sejnowski: What I meant was that we need a standard: some official body or sufficiently smart group of people who can anticipate all the different things that will be needed in the future and create a framework that will grow with these needs. The problem all along has been that when one person decides to do this it usually ends up becoming obsolete very quickly.

Brenner: That is why I think the Linux precedent is very good. Someone started it, and because it was open source, people contributed to it and it works.

Noble: That is the operating system. Even with that it should not matter whether people code in C, Fortran or Pascal. This is what I mean when I say that we should be able to communicate.

Sejnowski: There are computer scientists who are trying to develop interfaces between tools. You want to be able to define the variables that are going back and forth between the programs, regardless of how they are implemented. It is not a trivial problem.

Laughlin: There is one thing that we haven't discussed which I think is central to scientific investigation, the hypothesis. We have seen that we have a wide variety of available models and we have systems that are undetermined. The first thing that we have to do is to generate hypotheses. You think that it is one sort of system, a richly interconnected diffuse system, and someone else thinks that it is a different sort: a small network of proteins that forms a well defined computational module. Proceeding from these hypotheses you construct the appropriate models and see whether they work. If a model doesn't work you throw it away and try to think of a better one. This is a very important part of modelling. We make decisions a priori about what sort of system it is, and then construct a model on that basis to see whether it works. You don't sit there and wait until you have all the information.

Segel: On the question of levels of models, my opinion is that you will never be able to link the levels fully. My opinion stems from the situation in a simple subject, physics. In fluid mechanics, in the 19th century scientists derived the Navier–Stokes equations for fluid flow. For very dilute gases, you can deduce the dependence of viscosity on temperature and pressure, but for something like water, it is so complicated that we are a long way off being able to do this. So what do people do? They measure the viscosity dependence and thereby operate successfully at the continuum level for a certain class of questions. For other classes of questions they have to go to other levels. Conceptually one can dimly see the links between the models, but practically, we have to operate separately at different levels.

Brenner: A physicist looking at a complex system takes a very different stance from that of a biologist. Biologists understand the overall constraint imposed by the fact of evolution; we are in the business of trying to explain a very complex system that arose in a self-organizing way and carries with it the marks of its history.

Von Neumann made a remark many years ago in which he said that when things reach a high level of complexity, maybe one ceases to give a theoretical explanation of it, possibly because there is no meta-language to call on to couch the explanation. The alternative he proposed was to create a device that generates that behaviour. Today we would call this an algorithmic explanation.

Sejnowski: That sounds like a justification for artificial intelligence. This relates to the issue of levels. This grand experiment was done — it started around 1956 — and it is just about over and the returns are in. If you go about it in this way, you are almost bound to fail. It is interesting to look at why it failed. In retrospect, researchers failed to appreciate how a formalism can affect your way of thinking. At the time that they started, digital computers were really slow (a few thousand operations per second), and the one thing that the digital computer did exceedingly well (because of the machine language) was the ability to manipulate symbols, because a symbol only takes one bit to code. Furthermore, Turing had proven that anything that can possibly be computed can be done with that architecture — therefore we should be able to write a program that is intelligent or can see, or indeed have any other property that is computable. They set about trying to do this and ended up going down one blind alley after another. In retrospect, what happened is that if you constrain yourself to the part of computational space that is purely described by Boolean expressions, it may be that the Boolean expression that describes intelligence is so complicated that no one will be able ever to write it down. It is a problem of discovery. How is our own brain able to conceptualize and formulize complex systems? Recently, a new approach is being taken which starts with a mathematical formalism that looks more like the brain itself, namely neural networks. There are many simplifications and abstractions, starting with the assumption that there are massively parallel operations, and that the fundamental operation is arithmetic rather than Boolean. It is also a limited model, in the sense that you have now put yourself into another part of computational space, and it is a framework that may or may not be more conducive for us to discover general computational principles. The point is, you have to be very aware that when you start with any mathematical formalism that it will bias your thinking, in terms of what is easy to express in that formalism and what is difficult.

Berridge: I'd like to lead us back to where we started from with Denis Noble's comment about William Harvey taking a reductionist's approach to analysing the eel's heart. As experimentalists, we have usually gone top–down. We have started with the intact heart and tried to understand it, working out where the pacemaker is, where the atrial cells are, what channels are expressed and so on. Now we are arguing about whether or not this is a reasonable approach and Simon Laughlin has said that we need a good hypothesis. I would argue that Denis Noble already has a very good hypothesis about how the heart works. I would also argue that Denis has started at the molecular level. For example, he has begun with the channels and has modelled the action potential of a ventricular cell. Then, of course, the problem is to describe how such cells interact with each other in the intact heart, as Denis Noble has described. Unlike the experimental approach, which is top–down, it seems that modelling has to go from the bottom–up.

Noble: I would add that the choice of the formalism for each of those proteins was determined by what they were going to be used for in the physiological model. Let me give one example: when we modelled the Na^+/Ca^{2+} exchanger, we concentrated on getting the voltage and ion dependence of the flux of that exchanger right. At the same time, there was a group in Duke University publishing a minimal model of the Na^+/Ca^{2+} exchange. It represented around 40 reactions! It was never used by anyone else. This illustrates that there still has to be a functional viewpoint, and the maths that we chose constrained what we could then do with the model. This, in turn, was constrained by the physiology.

Sejnowski: This is something that is informing all of these choices: the physiology. This can occur at many different levels: the organ, the cell or even the channel. We have to be sensitive to what is really important. Casey Cole invented the voltage clamp, and taught Hodgkin how to use it. Cole had a theory for the action potential that was based on adding an inductance to the RC circuit. Once he started down this path, he ended up going in the wrong direction. Hodgkin and Huxley wrote down their kinetic equations not as a first choice, but as a fall-back, because they wanted to have a biophysical model. They ended up just fitting a bunch of equations which they felt was simply a phenomenological model, In retrospect, these equations fit very well with the kinetics of ionic channels. Two different models: one led down a dead end, the other to a highly successful description of excitable cells.

References

Mainen ZF, Sejnowski TJ 1995 Reliability of spike timing in neocortical neurons. Science 268:1503–1506

Reed MA, Tour JM 2000 Computing with molecules. Sci Am 275:69–75

Segel LA Grappling with complexity. Complexity 1:18–25

Regulation of gene expression by action potentials: dependence on complexity in cellular information processing

R. Douglas Fields, Feleke Eshete, Serena Dudek, Nesrin Ozsarac and Beth Stevens

National Institute of Child Health and Human Development, National Institutes of Health, Bethesda, MD, USA

Abstract. Nervous system development and plasticity are regulated by neural impulse activity, but it is not well understood how the pattern of action potential firing could regulate the expression of genes responsible for long-term adaptive responses in the nervous system. Studies on mouse sensory neurons in cell cultures equipped with stimulating electrodes show that specific genes can be regulated by different patterns of action potentials, and that the temporal dynamics of intracellular signalling cascades are critical in decoding and integrating information contained in the pattern of neural impulse activity. Functional consequences include effects on neurite outgrowth, cell adhesion, synaptic plasticity and axon–glial interactions. Signalling pathways involving Ca^{2+}, CaM KII, MAPK and CREB are particularly important in coupling action potential firing to the transcriptional regulation of both neurons and glia, and in the conversion of short-term to long-term memory. Action potentials activate multiple convergent and divergent pathways, and the complex network properties of intracellular signalling and transcriptional regulatory mechanisms contribute to spike frequency decoding.

2001 Complexity in biological information processing. Wiley, Chichester (Novartis Foundation Symposium 239) p 160–176

Development and plasticity of the nervous system are regulated by neural impulse activity, which encodes information in the temporal pattern of action potentials. If persistent changes in nervous system structure and function are to result from neural impulse activity, regulation of gene expression is likely to be involved. This implies that gene expression must be regulated by the temporal pattern of action potential firing, not simply the presence or absence of activity. In contrast to what is known about temporal coding in neural circuits, much less is known of how information contained in the temporal pattern of membrane depolarization is transduced and integrated within the neuron to produce an appropriate transcriptional response in a specific gene.

FIG. 1.   Preparation for studying the effect of action potential firing patterns on intracellular signalling and gene expression in mammalian neurons. DRG neurons from fetal mice are cultured in multicompartment cell culture chambers equipped with platinum electrodes for electrical stimulation (Fields et al 1992). Neurons are stimulated to fire action potentials at different frequencies and in different patterns, and Ca^{2+} imaging and molecular techniques are used to monitor responses in signalling pathways and gene expression.

Our aim has been to explore the hypothesis that transcription of specific genes could be regulated by the temporal pattern of action potential firing. In pursuing this question, we were interested in identifying genes that could have important structural and functional effects on the nervous system. Secondly, we wished to determine whether structural and functional responses relevant to nervous system development and plasticity could be regulated by appropriate patterns of action potential firing. Finally, intracellular signalling mechanisms that could activate specific genes by appropriate action potential firing patterns were investigated.

To investigate these questions, we developed a biological model system that allows control of the input function (action potential firing pattern) and a method to monitor the output function (second messenger dynamics, kinase activity and gene expression) (Fields et al 1992). This is a preparation of mouse dorsal root ganglion (DRG) neurons cultured in a multicompartment cell culture chamber equipped with electrodes for extracellular stimulation (Fig. 1). These neurons offer several advantages for these studies: (1) they do not form synapses in culture, (2) they lack spontaneous impulse activity, and (3) they respond to brief electrical stimulation with a single action potential. These features allow the frequency and pattern of action potential firing to be controlled precisely and indefinitely in the incubator. Furthermore, direct membrane depolarization activates a much simpler intracellular signalling cascade than would be evoked by synaptic stimulation, with the attendant release of neurotransmitters and neuromodulators in a synaptic network of excitatory and inhibitory connections.

Using this preparation we have demonstrated functional effects of action potentials on a wide range of developmental phenomena, including growth cone motility (Fields et al 1990, 1993), synaptic plasticity (Nelson et al 1989, Fields et al 1991), Ca^{2+} channel expression (Li et al 1996), cell–cell adhesion (Itoh et al 1995, 1997), axon fasciculation and defasciculation (Itoh et al 1995), neuron–glial interactions (Stevens & Fields 2000), myelination (Stevens et al 1998) and cell proliferation (Stevens & Fields 2000). In each case, the frequency and pattern of action potentials have been investigated, and in most cases the pattern of action potential firing is an important factor in the functional response.

Regulation of c-*fos* by action potential firing pattern

The immediate early gene c-*fos* codes a transcription factor that binds to AP-1 recognition sites in the promoter region of other genes, and thereby regulates long-term adaptive responses to relatively brief physiological stimuli (Sheng et al 1990). Transcription of this gene is rapidly induced and the biochemical and molecular mechanisms regulating its transcription are understood in detail. By means of a combination of electrophysiology, Ca^{2+} imaging and molecular biological methods, the stimulus transcription pathway has been traced from action potentials through Ca^{2+} fluxes, to the protein kinases and transcription factors regulating transcription of the c-*fos* gene. The temporal dynamics of steps along the signalling pathway have been quantified, and the extent to which temporal dynamics, spatial heterogeneity in Ca^{2+} signalling or levels of cytoplasmic Ca^{2+} can confer specificity between stimulus and response has been examined. The results emphasized the importance of the dynamic properties of intracellular signalling cascades working as a complex system to provide specificity between stimulus and response.

Expression of c-*fos* increases as a non-monotonic function of action potential frequency (Sheng et al 1993). Surprisingly low frequencies of action potentials are sufficient to induce transcription of this gene. Expression of c-*fos* increases significantly in response to one action potential delivered every 10 s (0.1 Hz) for 30 min. Near maximal levels of c-*fos* mRNA are reached after delivering 1 action potential per second (1 Hz), with minimal further increase in response to higher frequency stimulation.

Delivering the same number of action potentials (180) in a 30 min period results in very different levels of c-*fos* expression, depending upon how bursts of action potentials are patterned over time (Fig. 2). Delivering 180 action potentials in a single 18 s burst at 10 Hz was not effective in stimulating transcription, but delivering regular short 0.6 s bursts (6 action potentials at 10 Hz) every minute was more effective than constant frequency stimulation at 0.1 Hz (Fig. 2). However, longer bursts of action potentials at 10 Hz (1.2 s or 12 action potentials/burst) delivered at longer inter-burst intervals (every 2 min), was also ineffective in stimulating transcription of this gene. Interestingly, a different gene, *nur77* showed a different pattern of sensitivity to these stimuli (Sheng et al 1993). The results indicate that individual genes can be regulated by different patterns of action potentials, and that the temporal pattern of action potentials can be a more critical factor than the amount of stimulation (number of action potentials in a burst or the total number of action potentials delivered).

The signal transduction cascade by which membrane depolarization activates c-*fos* transcription has been characterized in detail. Depolarization causes an influx of extracellular Ca^{2+}, which activates Ca^{2+}-dependent kinases that carry the signals to the nucleus via a cascade of reactions ending in the phosphorylation of transcription factors bound to regulatory sequences in the promoter region of the gene (Fig. 3). Differences in intracellular Ca^{2+} concentration produced by different action potential firing patterns might provide specificity in signalling by activating appropriate enzymes based on their individual Ca^{2+} affinities. However, a simple relation between the concentration of intracellular Ca^{2+} and gene expression appears unable to account for the response of c-*fos* to these different action potential patterns. The results of Ca^{2+} imaging in neurons stimulated with these action potential patterns showed that gene expression was not dependent on a sustained increase in intracellular Ca^{2+} (Sheng et al 1993). All the stimuli produced only a transient increase in intracellular Ca^{2+} that returned to baseline within a few seconds of a stimulus burst. Secondly, gene expression in this case is not easily explained by the hypothesis that transcription depends upon reaching a particular Ca^{2+} concentration threshold, because a large increase in intracellular Ca^{2+} was not necessary to stimulate gene transcription, and in some instances was ineffective in doing so (e.g. bursts of 12 action potentials repeated at 2 min intervals). Conversely, the smallest Ca^{2+} transient that could be induced — that accompanying a single action potential (which produces only a 20 nM brief Ca^{2+} transient) — was effective in activating transcription when repeated at 10 s intervals. In contrast, two stimulus patterns producing the largest increase in Ca^{2+} (bursts of 12 action potentials repeated at 2 min intervals, or 180 action

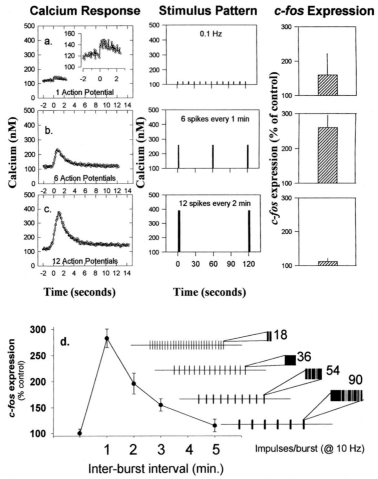

FIG. 2. The relationship between action potential firing pattern, intracellular Ca^{2+} concentration, and levels of mRNA for the gene c-*fos* was investigated in DRG neurons in multicompartment cell cultures. When neurons were stimulated to fire 180 action potentials in a 30 min period, we found that short bursts of action potentials every minute (b) were more effective in activating c-*fos* transcription than the same number of action potentials delivered in longer bursts, but repeated at 2 minute intervals (c). Fura-2 measurements of the intracellular Ca^{2+} transients evoked by action potential firing in these different patterns, show that neither high levels of intracellular Ca^{2+} nor prolonged elevation of intracellular Ca^{2+} are necessary to activate c-*fos* transcription. Extremely low frequency stimulation (0.1 Hz) produces a significant increase in c-*fos* expression (a) although this produces only a minimal change in intracellular Ca^{2+} concentration. A similar relation holds for c-*fos* expression in response to 540 action potentials grouped in longer duration bursts (d). The results suggest that the temporal dynamics of second messenger generation, not only peak or steady-state concentration levels, are critical in understanding how neural impulse firing patterns activate transcription of specific genes. Adapted from Fields (1996) and Fields et al (1997) with permission.

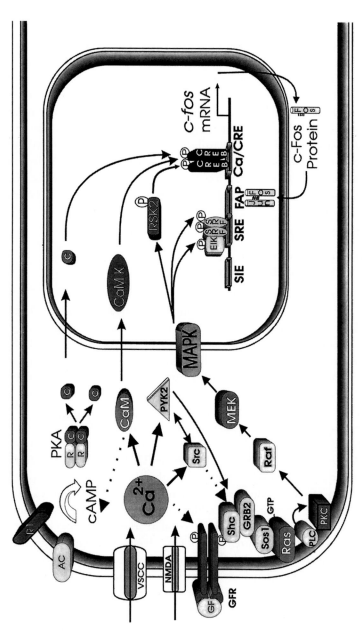

FIG. 3.   Multiple Ca^{2+}-activated signalling pathways with many points of interaction control transcription of the immediate early gene c-*fos* in response to membrane depolarization. The complex network behaviour of signalling pathways provides specificity between appropriate temporal patterns of action potentials and activation of specific genes. Intracellular signals from action potential firing propagate selectively through signalling pathways with temporal dynamics of activation and inactivation in neurons and glia in response to action potentials are Ca^{2+}, CaM depolarization, that are compatible with the temporal pattern of membrane Kinases (CaM K), MAP kinases (MAPK), and the Ca^{2+}/cAMP responsive element binding protein (CREB).

potentials at 10 Hz for 18 s), were not as effective as short bursts of action potentials (0.6 s) repeated at 1 min intervals.

Expression of c-*fos* was inversely correlated with the interval of time between bursts of action potentials, and this has been confirmed in a larger series of stimulus patterns using longer duration bursts that raise Ca^{2+} to relatively equal peak levels (1.8–9 s bursts, repeated at 1–5 min intervals) (Fig. 2) (Fields et al 1997). The results indicate that gene expression is not necessarily a function of the Ca^{2+} concentration dynamics; the temporal dynamics of intracellular Ca^{2+} concentration are an important factor. This conclusion is supported by similar findings in non-neuronal cells, where the temporal dynamics of Ca^{2+} oscillations (Dolmetsch et al 1998, Li et al 1998) or pulsatile hormone secretion have important functional consequences (Novartis Foundation 2000, Schofl et al 1993). Studies on frog spinal cord neurons during development show a relation between the frequency of spontaneous Ca^{2+} transients and maturation of K^+ channels (Gu & Spitzer 1995).

CaMKII autophosphorylation as a function of action potential frequency

The next step in signalling from action potentials is the activation of Ca^{2+}-dependent protein kinases, such as Ca^{2+}–calmodulin-activated protein kinase type II (CaMKII). It has been proposed that this enzyme could provide a mechanism for encoding Ca^{2+} spike frequency (Hanson et al 1994). High-frequency stimulation promotes autophosphoryaltion of CaMKII at Thr286 and subsequent CaM trapping (Hanson et al 1994), which sustains activation of the enzyme in the absence of Ca^{2+}. This could decode different action potential firing rates into different levels of sustained kinase activity (Meyer et al 1992). Experiments on purified alpha-CaMKII subjected to pulses of Ca^{2+}-containing solutions *in vitro* have demonstrated this behaviour and revealed several factors that modulate the frequency response of the enzyme (De Koninck & Schulman 1998). Among these are the duration of the individual Ca^{2+} pulses, the level of autophosphorylation of the enzyme prior to stimulation with Ca^{2+} pulses, and the particular isoform of the enzyme. In addition to these factors, phosphatase activity inside cells would be expected to regulate the frequency response (Dupont & Goldbeter 1992). Measurements of Ca^{2+} dynamics induced by action potentials in DRG neurons indicate that Ca^{2+} transients are much longer than those that can result in effective frequency-dependent activation of the enzyme (Eshete & Fields 1999). In addition, the isoforms of CaMKII in these cells, and the high level of autophosphorylation of the enzyme prior to stimulation, predict that the enzyme would be unable to decode action potential frequencies of greater than 1 Hz. Direct measurements of CaMKII autonomous activity and

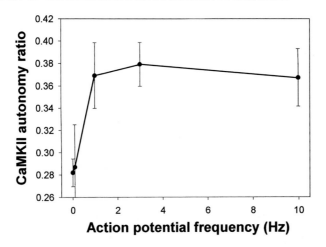

Action potential frequency (Hz)

FIG. 4.   Action potential firing patterns may be decoded by the autonomous activation of CaMKII resulting from autophosphorylation of the enzyme at Thr286 (De Koninck & Schulman 1998). However, measurements of autonomous activity of CaMKII in DRG neurons, indicate that this enzyme fails to decode firing patterns at frequencies of greater than 1/s. This would be expected from the high level of autophosphorylation in the basal state, the prolonged duration of the Ca^{2+} transients accompanying action potentials, and the predominant isoforms of CaMKII in these neurons (Eshete & Fields 1999).

autophosphorylation confirm this prediction (Fig. 4) (Eshete & Fields 1999). CaMKII autonomous activity may well decode action potential or intracellular Ca^{2+} pulse frequencies in other neurons or in specialized subcellular compartments, but this mechanism does not fully account for the frequency-dependent response in DRG neurons.

Spatial/temporal segregation in signalling networks

Rather than dependence upon a single signalling molecule, these data suggest that the network of intracellular signalling pathways may operate as a complex system to provide specificity between temporal patterns of stimulation and appropriate cellular response (Fields 1996). In addition to CaMKII, MAPK is activated in DRG neurons in response to action potential-induced Ca^{2+} influx. Both of these kinase cascades can act on transcription factors regulating c-*fos* transcription (Sheng et al 1991). Our measurements indicate that CREB is phosphorylated in DRG neurons by brief action potential stimulation (10 s at 10 Hz), and it remains activated for prolonged periods (tens of minutes) (Fields et al 1997). These kinetics would make CREB an excellent molecule for sustaining activation between bursts of action potentials separated by long inter-burst intervals, but would make it a

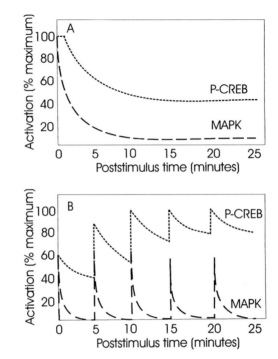

FIG. 5. Differences in inactivation kinetics of CREB and MAPK result in differential activation of these two signalling pathways in response to patterned action potential stimulation. (A) The inactivation kinetics of MAPK are much faster than the inactivation kinetics of CREB. (B) This allows temporal summation of CREB, but not MAPK in response to bursts of action potentials repeated at 3–5 min intervals. Neuronal responses and transcription factors dependent on activation of MAPK would not be activated effectively in response to this stimulus pattern. Adapted from Fields et al (1997).

poor enzyme for responding to different frequencies of action potentials, since it could not follow rapid firing rates with high fidelity. The kinetics of MAPK differ from CREB in that it is inactivated comparatively quickly, returning to near pre-stimulus levels within 3–5 min (Fig. 5A). This implies that bursts of action potentials arriving at intervals longer than 3–5 min would fail to propagate effectively through signalling pathways involving MAPK, and our measurements in intact DRG neurons corroborate this (Fig. 5B) (Fields et al 1997).

Ca^{2+}-dependent signalling from action potentials in hippocampus and glia

These intracellular signalling pathways are activated by action potentials in many other cells in the nervous system, including the means by which action potentials signal to peripheral glia (Stevens & Fields 2000), and the conversion of short-term to long-term memory (review see Dudek & Fields 1999). Using time-lapse laser scanning confocal microscopy, we observed Ca^{2+} responses in cultured Schwann cells when premyelinated DRG axons were stimulated electrically (Stevens & Fields 2000). This axon–Schwann cell communication is mediated by the non-synaptic release of ATP from DRG axons activating P2Y receptors on Schwann cells, and causing release of Ca^{2+} from intracellular stores. Further work showed that action potential firing activated these same Ca^{2+}-dependent signalling pathways in the Schwann cells, including CaMKII and MAPK, to phosphorylate CREB and stimulate transcription of the genes c-*fos* and *krox24*. The response varies directly with action potential frequency, and action potential firing rates as low as 1/s were effective. The effect of this axon–Schwann cell signalling was to regulate differentiation and proliferation of the Schwann cells (Stevens & Fields 2000).

The gene product responsible for this conversion of short-term to long-term memory is unknown, but work from several laboratories have implicated Ca^{2+}, CaMKII (Mayford et al 1995), MAPK (English & Sweatt 1997) and CREB phosphorylation (Bourtchuladze et al 1994, Yin et al 1995) in the signalling pathway activating transcription of the gene or genes responsible for long-term memory. It is not known, however, how these signals reach the nucleus (Frey & Morris 1997, Dudek & Fields 1999). A number of signalling molecules have been proposed that could carry signals from the subsynaptic membrane to the nucleus, including calmodulin (Deisseroth et al 1998), Ca^{2+}, BDNF and NF-κB (review by Suzuki 1996). We find that by blocking all excitatory synaptic transmission and inducing somatic action potentials in CA1 hippocampal neurons, that CREB phosphorylation and transcription of a gene associated with induction of long-term potentiation (LTP), *zif268*, could be stimulated by somatic action potentials without the need for a synapse–nucleus signalling molecule (Dudek & Fields 1998). This suggests that Ca^{2+} influx through somatic voltage-sensitive Ca^{2+} channels can trigger CREB-dependent gene expression necessary for long-term memory.

MAPK in these neurons could be activated by either LTP-inducing stimulation (theta-bursts at high frequency) or low frequency stimulation (3–10 Hz), which is not typically effective in inducing LTP. However, different Ca^{2+}-dependent signalling pathways were activated by these two stimulus paradigms. Ca^{2+} influx through the L-type Ca^{2+} channels or postsynaptic NMDA receptors is sufficient to allow phosphorylation of MAPK by LTP-inducing stimuli, but Ca^{2+} influx through NMDA receptors alone was primarily responsible for MAPK activation in response to the low-frequency stimulus protocol (Dudek & Fields 2001). This indicates that the frequency of synaptic input to postsynaptic CA1 hippocampal neurons is temporally and spatially resolved to activate distinct intracellular signalling pathways known to regulate gene expression and synaptic plasticity.

Late-response genes and functional effects
of patterned action potential firing

The use of gene arrays for expression profiling is revealing that a large number of genes can be expressed in response to a single stimulus, such as growth factor stimulation (Fanbrough et al 1999), and our preliminary observations in DRG neurons in response to action potential stimulation are consistent with this. Thus, the complex network behaviour that contributes to specificity in cytoplasmic signalling enzymes (Bhalla & Iyengar 1999, Weng et al 1999) appears to operate in transcriptional regulation in the nucleus as well (Robertson et al 1995, McAdams & Shapiro 1995). Some of the genes activated by appropriate action potential firing patterns would be expected to have major effects on nervous system development and plasticity. Expression of the cell adhesion molecule L1 is down-regulated in DRG neurons by 0.1 Hz stimulation, but 1 Hz stimulation is without effect (Itoh et al 1995). In contrast, N-cadherin mRNA is down-regulated by both 0.1 and 1 Hz stimulation, whereas NCAM-180 expression is not altered (Itoh et al 1997). These changes are associated with functional effects on the adhesion of axons into bundles (Itoh et al 1995). This could be important in regulating axon pathfinding and synaptogenesis in late stages of development.

Conclusion

Intracellular signalling cascades are characterized by multiple points of interaction, including cross-talk, convergence, divergence, synergism etc. This research *in vitro* supports the view that these points of interaction are necessary to allow intracellular signalling to function as a network. Differences in temporal dynamics of activation and inactivation of different signalling pathways allow appropriate patterns of temporally varying stimulation from action potentials to propagate preferentially through signalling pathways with favourable kinetics, thus providing one mechanism for stimulus–response specificity. Spatial heterogeneity in signalling reactions within neurons provides another important mechanism for stimulus–response specificity (e.g. Hardingham et al 1997). Differences in concentration of second messengers can be less important than the temporal dynamics when the cell is not in a saturating, steady-state stimulus condition, such as the normal physiological state. Rather than dependence on second messenger concentration or the properties of a single signalling molecule, the complex network properties of signalling reactions and transcriptional responses provide specificity and resiliency of neuronal responses to changing conditions and environments.

References

Bhalla US, Iyengar R 1999 Emergent properties of networks of biological signaling pathways. Science 283:381–387

Bourtchuladze R, Frenguelli B, Blendy J, Cioffi D, Schutz G, Silva A J 1994 Deficient long-term memory in mice with a targeted mutation of the cAMP-responsive element-binding protein. Cell 79:59–68

De Koninck P, Schulman H 1998 Sensitivity of CaM kinase II to the frequency of Ca^{2+} oscillations. Science 279:227–230

Deisseroth K, Heist EK, Tsien RW 1998 Translocation of calmodulin to the nucleus supports CREB phosphorylation in hippocampal neurons. Nature 392:198–202

Dolmetsch R, Xu K, Lewis R 1998 Calcium oscillations increase the efficiency and specificity of gene expression. Nature 392:933–936

Dudek SM, Fields RD 1998 Somatic action potentials are sufficient for rescue of tagged synapses. Soc Neurosci Abstr 24:1074

Dudek SM, Fields RD 1999 Gene expression in hippocampal long-term potentiation. Neuroscientist 5:275–279

Dudek SM, Fields RD 2001 Mitogen-activated protein kinase/extracellular signal-regulated kinase activation in somatodendritic compartments: roles of action potentials, frequency, and mode of calcium entry. J Neurosci 21:RC122

Dupont G, Goldbeter A 1992 Protein phosphoryaltion driven by intracellular calcium oscillations: a kinetic analysis. Biophys Chem 42:257–270

English JD, Sweatt D 1997 A requirement for the mitogen-activated protein kinase cascade in hippocampal long term potentiation. J Biol Chem 272:19103–19106

Eshete F, Fields RD 1999 Spike frequency decoding and autonomous activation of CaM kinase II in DRG neurons. Soc Neurosci Abstr 25:1192

Fanbrough D, McClure K, Kazlauskas A, Lander ES 1999 Diverse signaling pathways activated by growth factor receptors induce broadly overlapping, rather than independent, sets of genes. Cell 97:727–741

Fields RD 1996 Signaling from neural impulses to genes. Neuroscientist 2:315–325

Fields RD, Neale EA, Nelson PG 1990 Effects of patterned electrical activity on neurite outgrowth from mouse sensory neurons. J Neurosci 10:2950–2964

Fields RD, Yu C, Nelson PG 1991 Calcium, network activity, and the role of NMDA channels in synaptic plasticity *in vitro*. J Neurosci 11:134–146

Fields RD, Yu C, Neale EA, Nelson PG 1992 Recording chambers in cell culture. In: Kettenmann H, Grantyn R (eds) Electrophysiological methods for *in vitro* studies in vertebrate neurobiology. Liss, New York, p 67–76

Fields RD, Guthrie PG, Russell JT, Kater SB, Malhotra BS, Nelson PG 1993 Accommodation of mouse DRG growth cones to electrically induced collapse: kinetic analysis of calcium transients and set-point theory. J Neurobiol 24:1080–1098

Fields RD, Eshete F, Stevens B, Itoh K 1997 Action potential-dependent regulation of gene expression: temporal specificity in Ca^{2+}, cAMP-responsive element binding proteins, and mitogen-activated protein kinase signaling. J Neurosci 17:7252–7266

Frey U, Morris RGM 1997 Synaptic tagging and long-term potentiation. Nature 385:533–536

Gu X, Spitzer NC 1995 Distinct aspects of neuronal differentiation are encoded by frequency of spontaneous Ca^{2+} transients. Nature 375:784–787

Hanson PI, Meyer T, Stryer L, Schulman H 1994 Dual role of calmodulin in autophosphorylation of multifunctional CaM kinase may underlie decoding of calcium signals. Neuron 12:943–956

Hardingham GE, Chawla S, Johnson CM, Bading H 1997 Distinct functions of nuclear and cytoplasmic calicum in the control of gene expression. Nature 385:260–265

Itoh K, Stevens B, Schachner M, Fields RD 1995 Regulated expression of the neural cell adhesion molecule L1 by specific patterns of neural impulses. Science 270:1369–1372

Itoh K, Ozaki M, Stevens B, Fields RD 1997 Activity-dependent regulation of N-cadherin in DRG neurons: differential regulation of N-cadherin, NCAM, and L1 by distinct patterns of action potentials. J Neurobiol 33:735–748

Li M, Jia M, Fields RD, Nelson PG 1996 Modulation of calcium currents by electrical activity. J Neurophysiol 76:2595–2607

Li WH, Llopis J, Whitney M, Zlokarnik G, Tsien RW 1998 Cell-permeant caged InsP$_3$ ester shows that Ca^{2+} spike frequency can optimize gene expression. Nature 392:936–941

Mayford M, Wang J, Kandel ER, O'Dell TJ 1995 CaMKII regulates the frequency-response function of hippocampal synapses for the production of both LTD and LTP. Cell 81:891–904

McAdams HH, Shapiro L 1995 Circuit simulation of genetic networks. Science 269:650–656

Meyer T, Hanson P, Stryer L, Schulman H 1992 Calmodulin trapping by calicum-calmodulin-dependent protein kinase. Science 256:1199–1202

Nelson PG, Yu C, Fields RD, Neale EA 1989 Synaptic connections in vitro: modulation of number and efficacy by electrical activity. Science 244:585–587

Novartis Foundation 2000 Mechanisms and biological significance of pulsatile hormone secretion. Wiley, Chichester (Novartis Found Symp 227)

Robertson LM, Kerppola TK, Vendrell M et al 1995 Regulation of c-fos expression in transgenic mice requires multiple interdependent transcriptional control elements. Neuron 14:241–252

Schofl C, Brabant G, Hesch RD, von zur Muhlen A, Cobbold PH, Cuthbertson KS 1993 Temporal patterns of alpha 1-receptor stimulation regulate amplitude and frequency of calcium transients. Am J Physiol 265:C1030–C1036

Sheng HZ, Fields RD, Nelson PG 1993 Specific regulation of immediate early genes by patterned neuronal activity. J Neurosci Res 35:459–467

Sheng M, McFadden G, Greenberg ME 1990 Membrane depolarization and calcium induce c-fos transcription via phosphorylation of transcription factor CREB. Neuron 4:571–582

Sheng M, Thompson MA, Greenberg ME 1991 CREB: a Ca^{2+}-regulated transcription factor phosphorylated by calmodulin-dependent kinases. Science 252:1427–1430

Stevens B, Fields RD 2000 Response of Schwann cells to action potentials in development. Science 287:2267–2271

Stevens B, Tanner S, Fields RD 1998 Control of myelination by specific patterns of neural impulses. J Neurosci 15:9303–9311

Suzuki T 1996 Messengers from the synapses to the nucleus (MSNs) that establish late phase of long-term potentiation (LTP) and long-term memory. Neurosci Res 25:1–6

Weng G, Bhalla US, Iyengar 1999 Complexity in biological signaling systems. Science 284:92–96

Yin JC, Del Vecchio M, Zhou H, Tully T 1995 CREB as a memory modulator: induced expression of a dCREB2 active isoform enhances long-term memory in Drosophila. Cell 7:107–115

DISCUSSION

Noble: Did I understand correctly that you said that in investigating the messaging from synapses to the nucleus, it was not necessary that there should be such transmission, because you can show that everything can be accounted for by Ca^{2+} signals that arise from full scale action potentials down the axon?

Fields: Yes.

Noble: I now want to unpack why you said this was not necessary but it might still happen. One reason why it might still happen is that you might want the specificity. This leads me to my question. The trouble with discharge of the whole axon is that this is an integrative activity of a large number of synapses. This doesn't discriminate between different inputs. If you want to have complexity that is to do with the nature of the input, you would want it to be the case that it matters which synapses have been activated. Was this the reason for your hesitation?

Fields: No. In fact, I don't think that is necessarily the case. The labelling specificity is determined by synaptic stimulation. Frey & Morris (1997) showed that even brief synaptic stimulation tags that synapse (The brief stimulation referred to here is one that is sufficient to induce e-LTP). But the gene product could be induced by somatic depolarization turning on CREB. This brings us back to what Hebb proposed originally in his rule for synaptic modification: that it was the firing of the postsynaptic neuron that provided the necessary condition for strengthening the coincidentally active synapse. It may be that short-term plasticity does not require a spike, but it may be that in the long-term, it is when the neuron fires that there is the appropriate condition to induce a permanent change in synaptic efficacy. The reason I hesitated was only that we don't yet have experimental evidence to exclude the possibility that late LTP (l-LTP) can occur without a somatic spike. It is very hard to exclude this, because it would require 10 h intracellular recordings.

Noble: But there could be interaction between changes in gene expression due to global activity and the preexisting synaptic changes due to specific synapses acting. Is that possible?

Fields: Yes.

Sejnowski: When you used antidromic stimulation, what was the pattern?

Fields: The pattern we used for antidromic stimulation was the same pattern delivered synaptically to induce l-LTP. We are trying to determine the pattern of antidromic stimulation necessary to convert e-LTP to l-LTP. In a way, the pattern of afferent stimulation required to induce LTP in experimental preparations is necessary to recruit massive activation of multiple synaptic inputs to raise the neuron above spike threshold. So it may not be necessary to make the cell body fire at θ burst patterns of action potentials.

Sejnowski: If you stimulate synaptically with θ bursts, you don't necessarily get a θ burst response in the action potentials, because you are recruiting inhibitory networks that hyperpolarize the cell. It is going to be complicated to understand the relationship of the presynaptic and postsynaptic signals and what they contribute.

Fields: There is a lot of work to be done, but you might also argue the extreme case, that we only need one somatic action potential, if that were enough to turn on CREB.

Iyengar: We are doing some work on genes that has led us to a real block. Using the LTP model, we have found that inhibition of MAP kinase blocks increases in CaMKII level. Then we went ahead and did a fancy experiment with the gene array. It turns out that a number of genes that have been previously described show up. The most interesting thing is that CaMK doesn't show up at all.

Sejnowski: This may be a way to understand the complexity of LTP. Now that we know that in normal LTP there are cascades of many different genes that are regulating each other, it is likely that a lot of the confusion in the LTP field may arise from different experimental procedures playing out the different combinations of reactions, leading to different outcomes.

Berridge: I want to unpack this a little further with regard to the role of CREB phosphorylation in setting the stage to consolidate these synapses. The idea is that you activate CREB and the transcriptional products start drifting out into the dendritic tree, and from what I understood you were trying to say, they go to those synapses that have been activated in order to perform some consolidation process. In effect, the cell is now in a sensitive state ready for consolidation. What happens if you now activate another set of synapses? Will they use the information from a previous simulation to facilitate the consolidation event? If you go back to Hebb, he was saying that what we really need is coincidence between the post-synaptic and pre-synaptic processes. This then provides a unique tag on that synapse. You have to think a little more deeply about exactly how the transcription event functions to consolidate those synapses that have witnessed the unique Hebbian event. I don't see this connection in what you have been saying. It seems to me that this is a critical factor that one has to bear in mind.

Fields: Hebb was very careful in what he said. He did say something to the extent that the necessary condition for increasing the efficacy of a neuronal connection was provided when the presynaptic neuron *repeatedly* and *persistently* took part in *firing* the postsynaptic cell. In other words, he didn't really define the temporal window. There could be an as yet undefined temporal window with regard to what is meant by 'coincidence' of pre- and postsynaptic activation for different types of synaptic changes in strength. This temporal window may be different for the induction of LTP versus the consolidation into long-term synaptic potentiation. Induction of e-LTP does not require a somatic action potential, but perhaps, as Hebb originally proposed and our data suggest, the long-lasting form of LTP might.

Iyengar: The Frey & Morris (1997) paper is quite important for that. If the MAP kinase starts off at the synapse that was activated, and is going to the nucleus to get CREB, along the way it can tag the tracks. Let's assume that CREB activates the expression and synthesis of TPA, and TPA needs to go to that activated synapse. Now you have a synapse tagged by MAPIC so that even though TPA is all over the soma you can track it in one direction. In this respect you could still have TPA accumulate in a dendrite-specific manner, if that is what MAP kinase does.

Berridge: Then how does this work in Doug Fields' experiment with the antidromic stimulation?

Iyengar: I don't know.

Fields: It is also interesting that you can have MAP kinase turned on with a stimulus that doesn't induce LTP.

Iyengar: Sure, isoprotenal alone will turn on MAP kinase very well in neurons, and cAMP does not induce LTP.

Sejnowski: It may be that more than one signal pathway needs to be activated.

Dolmetsch: I was under the impression that people could induce long-lasting LTP by just depolarizing a post-synaptic cell, without it firing an action potential. If you pair that with stimulation, this is sufficient to induce LTP. Is that correct?

Fields: You are correct with regard to the induction of LTP, but those experiments are short-term: certainly less than an hour.

Dolmetsch: I am trying to reconcile your results with a couple of papers from Dick Tsien's laboratory (Mermelstein et al 2000, Deisseroth et al 1996). He says that CREB is mostly phosphorylated by synaptic input and not by non-synaptic input, and the reason for this is that L-type channels which are required are highly tuned to the specific inputs. If you just fire action potentials, this doesn't allow enough time for L-type channels to open, whereas if you have EPSPs it does.

Fields: Dr Tsien's work was done in dissociated hippocampal cultures, where this is probably the case, but it is not true in slices.

Segel: Have you made a model that is a reasonable representation of the network that shows how the inter-beat interval could work?

Fields: We are beginning to work on mathematical models for this. It is probably due to the differences in kinetics of the CREB and MAP kinase pathways. The transcription factors that regulate c-*fos* involve both DNA regulatory elements that are preferentially activated by MAP kinase and CREB (the SRE and CRE). Robertson and colleagues in the Curran and Morgan labs have shown that a combination of all these transcription factors is necessary to bring about high levels of transcription of c-*fos* (Robertson et al 1995). It seems reasonable to us that those patterns of stimulation that bring about the maximal activation of both MAPK and CREB pathways will activate a combination of these DNA binding elements and lead to higher levels of transcription. Our results are compatible with this.

Segel: Is there a functional reason why you would try to build a device that works this way?

Fields: I think so. To regulate a genetic event in response to neural activity you would not want to have a system that is sensitive to minute-to-minute changes in neural activity. The genome can't respond on second or millisecond intervals.

Therefore it makes sense to have a system that looks at the timing of a burst, not just the frequency of action potentials or how much second messenger is generated, but how often the cell fires. The inter-burst interval is a better measure of the overall pattern of activity in a given period.

Berridge: I have another comment with regard to complexity. I am intrigued by the fact that a lot of memory consolidation takes place during sleep. We are always trying to associate the immediate synaptic events with gene activation. I don't know how strong the evidence is that you have to have both events occurring at the same time. It is possible that the synapses are tagged when the memory is acquired, but that the consolidation process may occur during sleep when the neuronal cells go off-line. Thus there may be a large temporal separation between memory acquisition and consolidation.

Sejnowski: This should be prefaced with the comment that no one has ever demonstrated that real learning and memory has been caused by LTP, except perhaps in the amygdala during fear conditioning (e.g. Maren 1999). In terms of behaviour, we know a few facts. One of them is that you can interfere with consolidation if you interfere with certain phases of sleep.

Berridge: This is why I am intrigued by the idea that these two events could be separated in time. In all our models we try to associate them, but this may not be the case.

Sejnowski: The problem with learning on-line is that the same cells that are busy processing sensory input and conditioning motor input are the ones that have ultimately to be reorganized. There are many reasons for why it would be useful to separate the period when structural changes occur from the period when learning takes place.

References

Mermelstein PG, Bito H, Deisseroth K, Tsien RW 2000 Critical dependence of cAMP response element-binding protein phosphorylation on L-type calcium channels supports a selective response to EPSPs in preference to action potentials. J Neurosci 20:266–273

Deisseroth K, Bito H, Tsien RW 1996 Signalling from synapse to nucleus: postsynaptic CREB phosphorylation during multiple forms of synaptic plasticity. Neuron 16:89–101

Frey U, Morris RG 1997 Synaptic tagging and long-term potentiation. Nature 385:533–536

Maren S 1999 Long-term potentiation in the amygdala: a mechanism for emotional learning and memory. Trends Neurosci 22:561–567

Robertson LM, Kerppola TK, Vendrell M et al 1995 Regulation of c-*fos* expression in transgenic mice requires multiple interdependent transcription control elements. Neuron 14:241–252

Efficiency and complexity in neural coding

Simon B. Laughlin

Department of Zoology, University of Cambridge, Downing Street, Cambridge CB2 3EJ, UK

Abstract. Neural coding in the retina and lamina of fly compound eyes is amenable to detailed anatomical, physiological and theoretical analysis. This approach shows how identified cell signalling systems are optimized to maximize the transmission of information. Optimization reveals three familiar constraints, noise, saturation and bandwidth, and shows how coding can minimize their effects. Experiments reveal a fourth constraint, metabolic cost, whose properties favour the distribution of information among multiple pathways. The advantages of distributed codes will be offset by increasing complexity and the build up of noise. The optimization of coding in fly retina suggests that both noise and complexity will be reduced by matching each step in the system's operations to the input signal, and to the logical requirements of the network's ultimate function, pattern processing. This line of argument suggests tightly organized networks, laid out so that information flows freely and independently, yet patterned so that the necessary contacts and transactions are made quickly and efficiently.

2001 Complexity in biological information processing. Wiley, Chichester (Novartis Foundation Symposium 239) p 177–192

Brains are animals' evolutionary responses to demands that are as basic to life as the requirement for energy; the need to collect, transmit, process and store information. Thus, in common with systems that structure and organize cells, a nervous system will recognize signals, detect patterns, co-ordinate inputs and outputs, generate patterns in space and time, and store information. Nervous systems elaborate and extend a set of molecular mechanisms for intra- and intercellular communication that enable living systems to organize components and achieve homeostasis. Given these similarities in mechanism and function, nervous systems must typify some aspects of complexity in biological signalling. Because nervous systems are specialised for speed, efficiency and wide-scale integration, some of the molecular and cellular constraints that determine signalling and computation in complex systems will be obvious. This facility is illustrated by work on fly retina.

Coding in the fly retina

The retinas of the blowfly *Calliphora* and the housefly *Musca* fulfil requirements that are necessary for a proper understanding of a signalling system: well-defined structures, well defined functions, and the ability to monitor signals accurately in intact systems, in space and time (Laughlin 1994). Each optical module (ommatidium) of the compound eye defines an image pixel. A well-defined group of photoreceptors and interneurons (Fig. 1) code achromatic contrast at each pixel (Hardie 1986, Strausfeld 1989). Photoreceptors R1–6 drive the two large monopolar cells (LMCs) via an array of 1320 chemical synapses (Nicol & Meinertzhagen 1982). As the eye moves across an image, the contrast in each pixel changes and this is represented by continuous changes in the membrane potential, first in the photoreceptor, then in the LMC (Fig. 1). The signal is enhanced as it passes from photoreceptor to LMC. High-pass filtering, a product of transient responses, removes the standing background signal. The fluctuations about the background represent objects of different contrast, and these fluctuations are amplified to fill the LMC response range (Fig. 1). Because we can make stable intracellular recordings from these cells in the intact animal, this straightforward coding process has been rigorously analysed. Experiment and theory have revealed fundamental principles of coding applicable to cell signalling systems.

Coding information

The six photoreceptors and two LMCs maximize the amount of information captured from photons and transmitted to the brain (Laughlin 1994). More bits means more grey levels per pixel and a better view for the fly. Bits quantify the amount that a fly can learn about the world (Rieke et al 1997). The information rate, I bits/s, is given by Shannon's formula for analogue signals

$$ I = \int_0^b \log_2 \left[1 + \frac{S(f)}{N(f)} \right] df $$

where $S(f)$ and $N(f)$ are the power spectra of signal and noise respectively, f is frequency, and the bandwidth, b, is the highest frequency coded (Shannon 1949). Note that the information rate increases when the response is faster (b increases) and when it is more reliable (the signal to noise ratio term, $S(f)/N(f)$, increases). To measure the bit rate, I, we drive photoreceptors and LMCs with random (Gaussian white noise) modulations in stimulus contrast (Fig. 1) that resemble the natural inputs measured with a model fly's eye (de Ruyter van Steveninck & Laughlin 1996a). The signals and noise recorded from photoreceptors and LMCs are sufficiently linear and Gaussian to satisfy the conditions of Shannon's formula. In daylight, a single photoreceptor codes 1000 bits/s. The LMC transmits more,

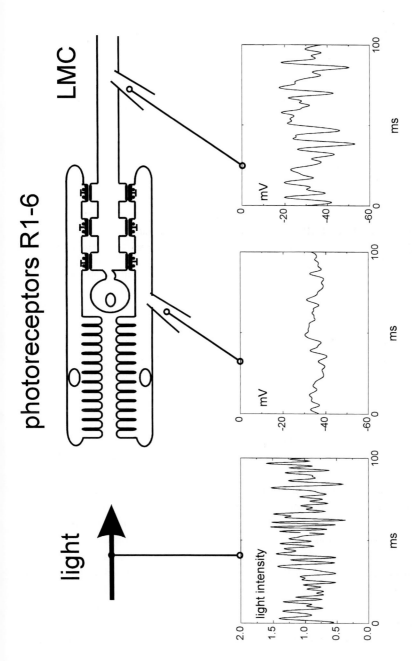

FIG. 1. The small module of photoreceptors R1–6 and large monopolar cells (LMCs) that code contrast at a single pixel of the fly retina. For clarity only two photoreceptors and one LMC are depicted. These cells respond to a modulation of light input with fluctuations in membrane potential that are recorded by intracellular micro-electrodes.

1600 bits/s, because it averages the signals from the six identical photoreceptors. From the number of synapses and the synaptic transfer function, we calculate that one chemical synapse transmits 55 bits/s. This is much more than the LMC bit rate divided by the number of synapses driving the LMC, because all synapses are carrying the same signal and this introduces massive redundancy. Averaging over the output of many synapses is a simple way of achieving a high bit rate in the face of synaptic noise. We presume that the fly uses this simple inefficient method because the efficient option, dividing the photoreceptor signal into independent components that can be passed through separate synapses, is too complicated to be implemented. The information rates in a photoreceptor and an LMC are two to three times those reported for axons transmitting action potentials (Rieke et al 1997), suggesting that simple and direct analogue codes can carry more information than pulsatile codes, at least over relatively short distances (de Ruyter van Steveninck & Laughlin 1996a).

Optimization identifies the constraints to signalling

The application of information theory to this system shows that coding is optimized to maximize bit rate (Laughlin 1981, van Hateren 1992a). Optimization confirms that the principal function of these cells is to maximise the number of bits acquired by the eye and that bit rate is a valid quantitative index of performance. The analysis of optimization defines the constraints that limit transmission, and demonstrates codes that minimize their effects. The constraints are noise, response range, bandwidth and metabolic cost.

Noise is ubiquitous because cell signalling involves diffusion and stochastic events (e.g. channel activation, molecular collision, the binding of ligand to receptor, vesicular release) (Berg 1983, Laughlin 1989). Noise can be reduced in a number of ways. Signalling complexes eliminate diffusion and chance collision, and force a stricter relationship between the number of events at the input (ligand bound) and at the output (enzymatic sites activated) (Bray 1998). Driving a system to saturation, with spikes, pulses or waves or by utilizing all the molecules in a restricted compartment, produces responses of constant amplitude. Signalling complexes (Montell 1998) and site saturation may contribute to the remarkable efficiency of blowfly photoreceptors, whose inositol-1,4,5-trisphosphate ($InsP_3$)-based phototransduction pathway (Hardie 1996) signals almost faultlessly to reach the limits set by photon noise (de Ruyter van Steveninck & Laughlin 1996b). In control systems, noise is reduced by negative feedback, as observed in motor systems and gene networks (Becskei & Serrano 2000). There is also the brute force method used by fly photoreceptors and LMCs, averaging over a population of components carrying the same signal. Finally, noise can be countered by boosting the signal but this raises the second constraint, saturation.

Saturation limits signal amplitude and power and, like signals, comes in many forms (reversal potential, maximum reaction rate, full occupancy, etc.). Because noise divides the signal into discriminable levels, saturation limits the number of signal levels, and hence the capacity to transmit information (note that I depends on $S(f)/N(f)$ in equation 1). Information theory formulates coding procedures that optimise coding by squeezing signals to fill this limited capacity. The two optimizations demonstrated by LMCs involve a precise match between coding and signal statistics.

The curve relating LMC response amplitude (mV) to stimulus strength (contrast), follows the cumulative probability function for contrast in natural scenes (Fig. 2). According to Information theory this match makes optimal use of the response range by ensuring that all response levels are used equally often (Laughlin 1981). Note that the LMCs contrast/response curve exhibits a non-linearity commonly observed in signalling systems and this coding strategy could be applied to receptors or enzymes by matching their dose-response curves to the statistical distribution of input signals.

The second optimal coding procedure adapts the LMC response waveform to the power spectra of natural inputs, as determined by natural image statistics and photon noise (van Hateren 1992a). At low light levels the response of an LMC is slow and monophasic to smooth out photon noise. As the light level increases the effect of photon noise reduces. The response waveform adjusts to remain optimum, becoming progressively faster and more biphasic (Fig. 2). By responding best to rapid and reliable changes, the LMC eliminates the redundancy generated by optics and correlations in natural images. Note that mechanisms that improve efficiency also increase the complexity of signalling, in this case by making the response waveform dependent on the previous history of input.

The third constraint is bandwidth, defined by the highest transmitted frequency, and equivalent to speed of response. LMCs make optimum use of bandwidth by speeding up their response as the input signal becomes more reliable (Fig. 2). Photoreceptors extend their bandwidth by reducing their membrane time constant. Time constant is the product of capacitance and resistance. Capacitance is generally fixed by membrane area, so to reduce their time constant photoreceptors reduce their membrane resistance by opening potassium channels (Weckström & Laughlin 1995). The resultant increase in transmembrane current introduces the fourth constraint, metabolic cost. This energy consumption can be equated with information.

The metabolic cost of information

Energy is required to generate the electrical signals of photoreceptors and LMCs. The major cost is providing metabolic energy to the pumps that restore

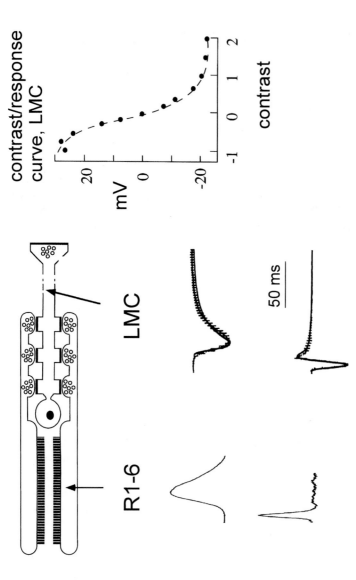

FIG. 2.   Two procedures that optimize coding in the blowfly compound eye. The contrast/response curve measured in a light-adapted LMC (data points) follows the statistical distribution of contrast in natural scenes (dashed line). Data from Laughlin et al (1987). The waveforms of responses to a brief flash of light, superimposed on a constant background, were recorded intracellularly from a photoreceptor R1-6 (S. B. Laughlin, unpublished data) and from LMCs (van Hateren 1992b). In the upper two traces the retina is weakly light adapted by a dim background. In the lower two it is strongly adapted by a bright background. The LMC waveforms compare the means and standard deviation of recorded responses (vertical lines) with the predictions of van Hateren's optimum coding model.

transmembrane ion fluxes. From biophysical measurements of ionic conductances and membrane potentials in intact cells we estimate ionic fluxes and derive the rate at which pumps consume ATP to maintain concentration gradients (Laughlin et al 1998). Dividing hydrolysis rate by bit rate gives a cost per bit of 7×10^6 ATPs in a photoreceptor. This makes vision expensive. Pumps in photoreceptors account for about 10% of oxygen consumption in the resting fly (Howard et al 1987). The cost in an LMC is lower, 2×10^6 ATP/bit as the result of redundancy reduction during coding, and photon noise reduction from the convergence of six photoreceptor signals. Calculations suggest that it would be just as expensive to transmit a bit through an LMC using an action potential. Clearly, we cannot automatically expect that pulsatile codes will be more economical than analogue ones.

A bit is over 20 times cheaper at a synapse. This suggests an important principle: the cost of a bit falls with bit rate (Laughlin et al 1998). Simple biophysics explains the dependence of bit cost on rate (Laughlin et al 1998). In an analogue systems, where signal amplitudes code information (e.g. membrane potential, ligand concentration or numbers of active sites), the information rate, I, increases with signal:noise ratio (SNR) and bandwidth (equation 1). To increase the SNR one must increase the number of events, molecules or organelles carrying the same signal. Because the SNR goes as the square root of number, and information goes as the log of the SNR (equation 1), the costs skyrocket as the rate increases (Fig. 3). In pulsatile signalling (spikes, Ca^{2+} transients, waves, etc.) the cost of a pulse is, to a first approximation, independent of rate, but cost will tend to increase slightly with rate because infrequent pulses can carry more information than frequent ones (Rieke et al 1997).

Metabolic cost and network complexity

Because unit costs increase with rate, it is more economical to transmit a certain number of bits (e.g. the bits that define the signal's accuracy and context) by distributing these bits among a number of weak channels of low capacity. Thus if the metabolic load of signalling is significant, distributed codes will be favoured (Laughlin et al 1998). Combinatorial considerations greatly increase the advantage of distributing information among weak channels. Consider 16 signalling particles (e.g. receptors, enzymes or second messenger molecules) that are subject to noise that goes as the square root of the number of particles. If all particles are used in the one pathway they code four states ($16/\sqrt{16}$ discriminable levels) but if the particles are distributed equally among four pathways they code 16 states (each pathway codes $4/\sqrt{4}=2$ levels, giving 2^4 combinations). The combinatorial advantage of distributed coding is illustrated by a fine analysis of metabolically efficient neural codes. This study demonstrates that the extent to which information should be distributed across pathways depends on the ratio between the cost of sending a

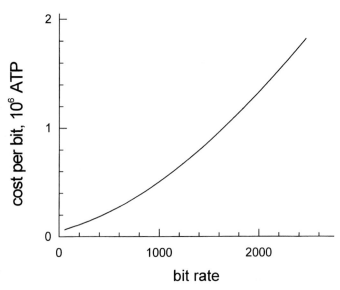

FIG. 3.   The metabolic cost per bit, expressed as ATP molecules hydrolysed, increases with the bit rate of a signalling pathway. The relationship was obtained by modelling a neuron that is driven by different numbers of identical synapses. Bit rate and cost rise with the number of synapses. Costs and rates are based on empirical estimates obtained for photoreceptor–LMC synapses in the fly compound eye (Laughlin et al 1998).

signal along a pathway and the cost of maintaining that pathway between signals (Levy & Baxter 1996).

Although economical, the division of signals among several pathways poses problems. An efficient system should avoid redundancy by dividing signals into independent components. As we have seen, a real system, such as the synapses from photoreceptors to LMCs, does not do this, presumably because it is too complicated. Where such division is possible the need for economy will enforce a logical set of coding processes that, by taking into account probabilities and context, will match coding to the structure of signals and patterns. Implementing this logical division must increase the number of operations and components and this will increase noise in the network. Perhaps the metabolic economies of distributed codes and the overheads of complicated networks strike a balance that defines the complexity of biologically relevant signalling systems?

Limits on complexity and stability imposed by noise can be countered by reducing noise. Noise accumulates (von Neumann 1958) and obscures signals in at least three situations, when different inputs converge, when signals pass through several stages and when signals cycle around loops (Fig. 4), as in models of neural motor pattern generators and circadian clocks where the build-up of noise

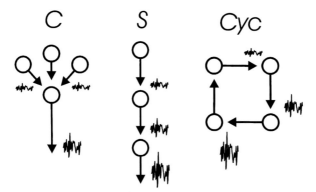

FIG. 4.   Noise builds up as signals pass through systems. Three configurations are illustrated, the convergence of signals on a single element (C), passage through a series of components (S), and passage around a cyclical system (Cyc).

seriously degrades performance (Miall 1989, Barkai & Leibler 2000). Pulsatile coding can, in conjunction with thresholding, eliminate noise (Sarpeshkar 1998, Laughlin et al 2000) but, as we have seen, spike codes transmit at lower rates than analogue. Hybrid analogue/digital systems offer the best of both worlds. The advantages of analogue computation — a rich set of primitives, flexibility, and high information rates — are exploited to perform small sets of computations locally. The accumulated noise is then stripped away by reducing the outcome of the analogue computation to a digital code whose pulses are faithfully transmitted as input to the next set of analogue processes. This hybrid system resembles cortical neurons where analogue signals are processed locally on dendrites and then integrated and transformed into an action potential train for transmission to the next stage (Sarpeshkar 1998). It will be interesting to see if this idea applies to cell signalling outside the nervous system (e.g. to Ca^{2+} transients that are triggered by small groups of signalling molecules).

Conclusions

We have identified three familiar constraints to cell signalling (noise, saturation and bandwidth), measured and modelled their effects, and discovered codes that optimize the processing of information within these constraints. These codes could well be applicable to a variety of cell signalling systems. Theoretical and experimental techniques are now being applied to a constraint that has previously received little formal analysis, the metabolic cost of information. Costs are high and scale with the amount of information to be transmitted so that a bit costs less in a communication channel of low information capacity than in a channel of high capacity. This trade-off between rate and cost promotes complexity by favouring the use of many low capacity channels; i.e. multiple signalling pathways. However, the other constraints, particularly noise, promote simplicity because these limitations are minimized by focusing signalling capacity on specific tasks. For example, the analysis of coding in the fly retina has shown that the effects of noise, saturation and bandwidth can be reduced by matching coding to the statistical properties of signals. Thus both noise and complexity will be reduced by matching each step in the system's operations to the input signal, and to the logical requirements of the network's ultimate function, pattern processing. Those logical operations that are common to different types of pattern processing could be encapsulated in signalling modules (Bray 1995, Hartwell et al 1999). Given a need for efficiency (e.g. from metabolic cost and crowding within the cell) these trade-offs could lead to tightly organized networks, laid out so that information flows freely and independently, yet patterned so that the necessary contacts and transactions are made quickly and efficiently. This scenario resembles the logical reliability of operations that von Neumann (1958) proposed as nature's solution to the problem of building a reliable brain from unreliable components. The design of such networks will be determined by the logic and statistics of the transactions being performed, and the mechanisms available from the genome. To understand their function we must have detailed descriptions of the processes being performed, the mechanisms being used and the statistics of signals. In physiological parlance, an understanding of integrative function requires that we link descriptions of anatomy and physiology to a detailed knowledge of both behaviour and the stimuli that determine behaviour. On this basis I suggest that we approach signalling complexity by considering pathways tied to specific sets of functions. This appears more promising than modelling the capabilities of an omnipotent spaghetti.

Acknowledgements

I gratefully acknowledge the support of the Gatsby Foundation, the Rank Prize Funds and the BBSRC.

References

Barkai N, Leibler S 2000 Circadian clocks limited by noise. Nature 403:267–268
Becskei A, Serrano L 2000 Engineering stability in gene networks by autoregulation. Nature 405:590–593
Berg HC 1983 Random walks in biology. Princeton University Press, Princeton
Bray D 1995 Protein molecules as computational elements in living cells. Nature 376:307–312
Bray D 1998 Signaling complexes: biophysical constraints on intracellular communication. Annu Rev Biophys Biomol Struc 27:59–75

de Ruyter van Steveninck RR, Laughlin SB 1996a The rate of information-transfer at graded-potential synapses. Nature 379:642–645

de Ruyter van Steveninck RR, Laughlin SB 1996b Light adaptation and reliability in blowfly photoreceptors. Int J Neural Systems 7:437–444

Hardie RC 1986 The photoreceptor array of the dipteran retina. Trends Neurosci 9:419–423

Hardie RC 1996 Calcium signaling: setting store by calcium channels. Curr Biol 6:1371–1373

Hartwell LH, Hopfield JJ, Leibler S, Murray AW 1999 From molecular to modular cell biology. Nature 402:C47–C52

Howard J, Blakeslee B, Laughlin SB 1987 The intracellular pupil mechanism and photoreceptor signal–noise ratios in the fly *Lucilia cuprina*. Proc R Soc Lond B Biol Sci 231:415–435

Laughlin SB 1981 A simple coding procedure enhances a neuron's information capacity. Z Naturforsch (C) 36:910–912

Laughlin SB 1989 The reliability of single neurons and circuit design: a case study. In: Durbin R, Miall C, Mitchison G (eds) The computing neuron. Addison-Wesley, New York, p 322–336

Laughlin SB 1994 Matching coding, circuits, cells, and molecules to signals: general principles of retinal design in the flys eye. Prog Ret Eye Res 13:165–196

Laughlin SB, Howard J, Blakeslee B 1987 Synaptic limitations to contrast coding in the retina of the blowfly *Calliphora*. Proc R Soc Lond B Biol Sci 231:437–467

Laughlin SB, de Ruyter van Steveninck RR, Anderson JC 1998 The metabolic cost of neural information. Nat Neurosci 1:36–41

Laughlin SB, Anderson JC, O'Carroll DC, de Ruyter van Steveninck RR 2000 Coding efficiency and the metabolic cost of sensory and neural information. In: Baddeley R, Hancock P, Foldiak P (eds) Information theory and the brain. Cambridge University Press, Cambridge, p 41–61

Levy WB, Baxter RA 1996 Energy-efficient neural codes. Neural Comput 8:531–543

Miall RC 1989 The diversity of neuronal properties. In: Durbin R, Miall RC, Mitchison G (eds) The computing neuron. Addison-Wesley, New York, p 11–34

Montell C 1998 TRP trapped in fly signaling web. Curr Opin Neurobiol 8:389–397

Nicol D, Meinertzhagen IA 1982 An analysis of the number and composition of the synaptic populations formed by photoreceptors of the fly. J Comp Neurol 207:29–44

Rieke F, Warland D, de Ruyter van Steveninck RD, Bialek W 1997 Spikes: exploring the neural code. MIT Press, Cambridge, MA

Sarpeshkar R 1998 Analog versus digital: extrapolating from electronics to neurobiology. Neural Comput 10:1601–1638

Shannon CE 1949 Communication in the presence of noise. Proc Inst Radio Eng 37:10–21

Strausfeld NJ 1989 Beneath the compound eye: neuroanatomical analysis and physiological correlates in the study of insect vision. In: Stavenga DG, Hardie RC (eds) Facets of vision. Springer-Verlag, Berlin, p 317–359

van Hateren JH 1992a Theoretical predictions of spatiotemporal receptive-fields of fly LMCs, and experimental validation. J Comp Physiol A 171:157–170

van Hateren JH 1992b Real and optimal neural images in early vision. Nature 360:68–70

von Neumann J 1958 The computer and the brain. Yale University Press, New Haven, CT

Weckström M, Laughlin SB 1995 Visual ecology and voltage-gated ion channels in insect photoreceptors. Trends Neurosci 18:17–21

DISCUSSION

Sejnowski: What is the difference between the amount of energy that is being consumed when you are thinking and when you are resting?

Laughlin: I recently learnt from Ritchie & Keynes (1985) that there is a fallacy here. If one thinks of ions traversing the membrane through channels, then the flow of this ionic current through these tiny resistors should heat the neuron. However Bernstein, working here in Berlin in the early 1900s, argued that this heating is counterbalanced by the cooling produced when ions pass from a compartment of high concentration to one of lower concentration. This cooling is equivalent to that produced during the adiabatic expansion of a gas. Thus there is no net heat production by current flowing across the neural membrane. Measurements of the heat produced by an action potential reveal two components. A small rapid pulse is associated with the change in membrane thickness induced by the change in potential, while a large slow component is generated by the activity of the Na^+/K^+ exchange pump (Ritchie & Keynes 1985).

With regard to the increase in metabolic rate in the active brain, my reading of the literature is that mental activity causes no perceptible increase in the overall oxygen consumption of the brain. We know from functional imaging studies, however, that there are local changes. But the people who do functional imaging do a lot of enhancement, and the increases are actually quite small, in the region of 10%.

Sejnowski: They also don't emphasize that when blood flow increases in one area, it decreases in others.

Laughlin: The other thing is that there is no significant change in oxygen consumption of the human brain during sleep. Whatever these theta rhythms are doing, they are generating the same levels of activity.

Sejnowski: During REM sleep, the metabolic rate is indistinguishable from that during the awake state. However, during slow-wave sleep the metabolic rate does go down somewhat.

Iyengar: If I can move completely away from neuronal systems, it seems that speed is not such an important thing. Is there an advantage in maintaining the reliability at low cost by simply slowing down? I started off studying desensitization back in 1978 in β-adrenergic or glucagon-stimulated systems. There, the stimulation and desensitization is very tightly coupled. The speed seems irrelevant. Are the other systems slowed down to increase reliability?

Laughlin: I think that the speed of signalling greatly increases the cost. I think the reason for this is twofold. In the nervous system you need shorter time constants in your neurons. This requires lower resistances, which make neurons leakier. Second, in any signalling system the signal must be turned on and then turned off again. If you need a fast signal, the mechanisms that turn the signal off have to work very effectively. They are probably going to chew up some of your signal before it even arrives at its destination. There is therefore some temporal overlap between the activation and inactivation mechanisms.

Iyengar: Does that make it more reliable?

Laughlin: Ultimately, the reliability depends on the total number of molecules that are being activated to transmit the signal, and the context. As we heard before, if you saturate your population of molecules, you will get a reliable signal. I think the dynamics of cell signalling are much more closely related to the temporal patterns that have to be generated within cells for them to produce appropriate responses.

Berridge: The process of desensitization is an interesting aspect. Were you measuring cAMP production and were you operating at agonist concentrations that are normally seen by a cell? Another important question is whether cells desensitize over their normal physiological range?

Iyengar: If you go back and look at John Perkins' papers, which are more careful than mine, there is a full overlap between maximal stimulation and desensitization (Su et al 1980). I don't remember the name of these partial agonists, but they had some that would stimulate but would not desensitize.

Berridge: But if you took a liver cell and studied glucose output, would you see a desensitization? These cells secrete for hours and we never see desensitization. As long as the dose is in the physiological range you won't see it; if you use a massive dose you see some strange things happening.

Iyengar: It may not be as fast as this, but in a few minutes glucose production shuts down, certainly with glucagon-sensitive cells.

Sejnowski: In digital computers, the amount of heat generated is directly related to the speed. There has to be a similar relationship.

Laughlin: That is presumably because the number of switches per second is going up in proportion to the speed. But I have been talking about the cost per switch (i.e. cost per pulse), and this is constant. An advantage of a pulsatile system over an analogue system is that when you increase the rate in a pulsatile system, the cost per bit tends to stay the same, but as we saw in Klaus Prank's paper (Prank et al 2001, this volume), the cost per bit goes up as you increase the rate because each spike can carry slightly less information because it is more frequent and therefore more predictable.

Sejnowski: Simon Laughlin and Klaus Prank, you raised the issue of information theory. This is a nice theory that is used by engineers in order to quantify how much information can be transmitted, for example, with cellular phones from a transmitter to a receiver. It is impressive that it agrees so well with experimental results. It is clear that the computation that needs to be done at the photoreceptor is to transduce the light signal into a form that is electrical and which can then be transmitted across that synapse. It makes sense that it should apply, and the fact that it does gives you some confidence both in your understanding of the system, but also in the ability of biology to meet the challenge of efficiency and speed of transmission. The question in my mind is how much of that formalism should apply to the next synapse. That is, is it the case that the purpose of the brain is simply to reproducibly transmit as much information as possible? In some respects, this would cause overload. If you had to deal with all the information that is coming into your brain, you would quickly be saturated with distractions. One of the important jobs the brain has is to sort out from all that information what is relevant for a particular task. This means that in some respects, one of the goals of the brain is to throw away information: to figure out which information is relevant and sort it out from what is not. This means that information theory by itself is probably not the only thing you need to apply in order to make these decisions.

Laughlin: Ultimately, what animals want to be able to do is to predict the most likely state of the world. They cannot establish the state of the world with total certainty; they predict. The accuracy of their predictions is a measure of the amount of information that they gain. This may be one way in which information theory can be developed further to understand behaviour and the brain, by determining the amount of information that can be inferred from the events in the world and comparing this with the ability of the brain to use that information. Terry Sejnowski is correct: information theory, as outlined here, is absolutely brilliant for determining the performance and the function of systems that are hungry for information, such as ears and eyes. Further up in the brain, what is important is semantics: what the information actually means. When Shannon developed this theory, he was most explicit. He states at the outset that he is not concerned with semantics. This immediately raises biologists' hackles, because the whole function of the nervous system is to determine which bits are the most important for survival. None the less, at higher levels of the nervous system, information theory tells you about 'good housekeeping'. It tells you that it is very wasteful to have a huge number of neurons, all carrying absolutely identical signals. This redundancy is a waste of space. It also gives you some idea about the representational capacity of different sorts of systems: how many states they can encode and what limits their ability to represent states. Note that if you don't have a unique state in your brain for something that is going on in the world, you can't discriminate it. Information theory gives us some clues as to what determines the number of unique states you can have.

Sejnowski: You are advocating something interesting, which relates to our discussion yesterday of levels. Here your whole theory can be applied to the level of the synapse, or the cell, which is a valid way of thinking. Then, at the level of semantics, you are suggesting that we should think about information theory but not at that level, but at the level of the representation of the whole pattern. In other words, the same theoretical foundations could apply at quite separate levels. My last challenge to you is the following. I am recording from an area deep in the cortex, and it produces an output spike train that typically has long intervals and short intervals. The membrane potentials are always fluctuating. How much of this is signal in terms of the timing of each impulse or spike, and how much of it is noise?

Laughlin: We don't know, because it seems highly likely that information is represented as patterns of activity distributed across cells. If you just look at the spike output of one cell, you learn very little about the pattern. What may be very important in this signal is the coincidence of a spike in this cell with a spike in another cell. I wouldn't try to infer very much from such a signal in a single cortical neuron.

Brenner: Information theory has been a source of confusion in my field. The way we use information when we talk about genetic information is not in the terms of information theory, but rather the way one uses 'information' when one talks about reading a newspaper for information. It is useful to think of this in terms of messages, rather than in terms of information.

Laughlin: I don't think information theory is the be all and end all of signalling. It happens to be the measure of useful work in the system I work with, the retina, and because it is fundamental it does offer us some insight into other systems. Everyone in neuroscience who is using information theory recognizes its limitations.

Dolmetsch: As we move up the evolutionary scale is the energetic optimization of coding lost to some extent? The reason I ask is that when I think of how photoreceptors work in mammals, it strikes me as highly inefficient in terms of energy use. They are continuously secreting neurotransmitter in the dark, and then they stop secretion when they encounter a photon. This is not an efficient design. When you become warm blooded, it costs a lot of energy just to be in the resting state. Perhaps in a moth it is essential that you consume the least amount of energy, but this constraint is lost further up the evolutionary tree.

Laughlin: As it happens, a moth is warm-blooded when active and this enables it to do more things per second. I believe that metabolic limitations could well promote the optimization of retinal function in vertebrates. Despite using different mechanisms, the cells of the vertebrate retina are subject to similar constraints. Although our rods are turned off in bright light by saturation, our cones have to continue to operate, scoring the arrival of every photon with the opening or closing of a channel. There has to be a coupling between one photon and at least one channel to code the signal. Consequently, the average number of channels open at any time must be roughly equal to the number of photons you transduce. This means that cones are no less or no more expensive to operate than insect photoreceptors. Moreover, our own retina shows the same sort of intensity-dependent coding that is demonstrated here in the insect. The fact that this optimum filtering of the input signal is used in both types of retina suggests that similar constraints operate in both cases.

Brenner: Years ago I tried to calculate the economy of *Escherichia coli* in terms of ATP. We looked at the cost of biosynthesis of all the components. We knew how much ATP *E. coli* made but we could only account for 10% of this in material synthesis. Later, it turned out that the cost of maintaining accuracy — all the repair and error correcting mechanisms — is much higher than the cost of actually making things.

References

Prank K, Kropp M, Brabant G 2001 Humoral coding and decoding. In: Complexity in biological information processing. Wiley, Chichester (Novartis Found Symp 239) p 96–110

Ritchie JM, Keynes RD 1985 The production and absorption of heat associated with electrical activity in nerve and electric organ. Quart Rev Biophys 18:451–476

Su YF, Harden TK, Perkins JP 1980 Catecholamine-specific desensitization of adenylate cyclase. Evidence for a multistep process. J Biol Chem 255:7410–7419

Neural dynamics in cortical networks — precision of joint-spiking events

Ad Aertsen*, Markus Diesmann†, Marc-Oliver Gewaltig‡, Sonja Grün¶ and Stefan Rotter*

*Neurobiology and Biophysics, Inst. Biology III, Albert-Ludwigs-University, Freiburg, †MPI Strömungsforschung, Department of Nonlinear Dynamics, Göttingen, ‡Future Technology Research, Honda R&D Europe, Offenbach, and ¶MPI Brain Research, Department of Neurophysiology, Frankfurt, Germany

Abstract. Electrophysiological studies of cortical function on the basis of multiple single-neuron recordings reveal neuronal interactions which depend on stimulus context and behavioural events. These interactions exhibit dynamics on different time scales, with time constants down to the millisecond range. Mechanisms underlying such dynamic organization of the cortical network were investigated by experimental and theoretical approaches. We review some recent results from these studies, concentrating on the occurrence of precise joint-spiking events in cortical activity, both in physiological and in model neural networks. These findings suggest that a combinatorial neural code, based on rapid associations of groups of neurons co-ordinating their activity at the single spike level, is biologically feasible.

2001 Complexity in biological information processing. Wiley, Chichester (Novartis Foundation Symposium 239) p 193–207

Modern attempts to understand the mechanisms of higher brain function are increasingly concerned with neuronal dynamics. The task of organizing perception and behaviour in a meaningful interaction with the external world prompts the brain to recruit its resources in a properly orchestrated manner. Contributions from many elements, ranging from individual nerve cells to entire brain areas, need to be coordinated in space and time. Our principal research goal is to understand how this organization is dynamically brought about, and how the brain uses such coordinated activity in neurons. To this end, we studied the spatiotemporal organization of cortical activity recorded at many different sites at a time. The rules that govern this organization and the underlying mechanisms are brought to light by complementary approaches of neurobiological experimentation, advanced data analysis, and neural network modelling.

According to the classical view, firing rates play a central role in neuronal coding (Barlow 1972, 1992). The firing rate approach indeed led to fundamental insights into the neuronal mechanisms of brain function (e.g. Georgopoulos et al 1993, Hubel & Wiesel 1977, Newsome et al 1989). In parallel, however, a different concept was developed, according to which the temporal organization of spike discharges within functional groups of neurons, the so-called neuronal assemblies (Hebb 1949), also contribute to neural coding (von der Malsburg 1981, Abeles 1982a, 1991, Gerstein et al 1989, Palm 1990, Singer 1993). It was argued that the biophysics of synaptic integration favours coincident presynaptic events over asynchronous events (Abeles 1982b, Softky & Koch 1993). Accordingly, synchronized spikes are considered as a property of neuronal signals which can indeed be detected and propagated by other neurons (Perkel & Bullock 1968, Johannesma et al 1986). In addition, these spike correlations must be expected to be dynamic, reflecting varying affiliations of the neurons depending on the stimulus or behavioural context. Such dynamic modulations of spike correlation at various levels of precision have in fact been observed in different cortical areas, namely visual (Eckhorn et al 1988, Gray et al 1989; for reviews see Engel et al 1992, Aertsen & Arndt 1993, Singer & Gray 1995, Roelfsema et al 1996), auditory (Ahissar et al 1992, Eggermont 1994, de Charms & Merzenich 1995, Sakurai 1996), somato-sensory (Nicolelis et al 1995), motor (Murthy & Fetz 1992, Sanes & Donoghue 1993), and frontal (Aertsen et al 1991, Abeles et al 1993a,b, Vaadia et al 1995, Prut et al 1998). Little is known, however, about the functional role of the detailed temporal organization in such signals.

The first important hints about the importance of accurate spike patterns came from the work of Abeles and colleagues (Abeles et al 1993a,b, Prut et al 1998). They observed that multiple single-neuron recordings from the frontal cortex of awake behaving monkeys contained an abundance of precise spike patterns. These patterns had a total duration of up to several hundred milliseconds and repeated with a precision of ± 1 ms. Moreover, these patterns occurred in systematic relation to sensory stimuli and behavioural events, indicating that these instances of precise spike timing play a functional role. Independent evidence for the possibility of precise spike timing in cortical neurons came from intracellular recordings *in vitro* (Mainen & Sejnowski 1995, Nowak et al 1997, Stevens & Zador 1998, Volgushev et al 1998) and *in vivo* (Azouz & Gray 1999).

We investigated the mechanisms underlying the dynamic organization of the cortical network by experimental and theoretical approaches. Here, we review evidence — both from experimental data and from model studies — that volleys of precisely synchronized spikes can propagate through the cortical network in a stable fashion, thereby serving as building blocks for spatiotemporal patterns of precisely timed spikes. Taken together, these findings support the hypothesis that precise synchronization of individual action potentials among groups of neurons presents an inherent mode of cortical network activity.

'Unitary events' in cortical multiple single-neuron activity

It has been proposed that cortical neurons organize dynamically into functional groups, so-called 'cell-assemblies' (Hebb 1949, Gerstein et al 1989). It is widely assumed that this functional organization is reflected in the temporal structure of the spike activity of the neurons involved. Thus, cortical activity would be characterized by synchronous spike volleys, travelling through the sparsely firing cortical network ('synfire chain' hypothesis; Abeles 1982a, 1991). To test this hypothesis, we analysed multiple single-neuron recordings from various cortical areas for the presence of excessive coincident spike events among the recorded neurons. We refer to such conspicuous coincidences as 'unitary events', and define them as those joint spike constellations that occur significantly more often than expected by chance (Grün et al 1994, Grün 1996). The functional significance of such unitary events was tested by investigating their occurrence and composition in relation to sensory stimuli and behavioural events.

'Unitary event' analysis

We developed a method that detects the presence of conspicuous spike coincidences and evaluates their statistical significance, taking into account the non-stationarities in the firing rates of the neurons involved (Grün 1996, Grün et al 2001a,b). Briefly, the detection algorithm works as follows: The simultaneous observation of spiking events from N neurons can be described mathematically by the joint process, composed of N parallel point processes. By appropriate binning, this can be transformed to an N-fold $(0,1)$-process, the statistics of which are described by the set of activity vectors reflecting the various $(0,1)$-constellations that occurred across the recorded neurons. Under the null-hypothesis of independently firing neurons, the expected number of occurrences of any activity vector and its probability distribution can be calculated analytically on the basis of the single neuron firing rates. The 'mutual dependence' measures the degree of deviation from independence among the neurons by comparing these theoretically derived probabilities with their empirical values. Those activity vectors that violate the null-hypothesis of independence define potentially interesting occurrences of joint events; their composition defines the set of neurons which are momentarily engaged in synchronous activity.

To test the significance of such unitary events, we developed a new statistical measure: the 'joint-P-value'. For any particular spike activity vector, this joint-P-value measures the cumulative probability of observing the actual number of coincidences (or an even larger one) by chance. Finally, in order to account for non-stationarities in the discharge rates of the observed neurons, modulations in spike rates and coincidence rates are determined on the basis of short data segments by sliding a fixed time window (typically 100 ms wide) along the data in steps of the coincidence binwidth. This timing segmentation is applied to each trial, and the data of corresponding segments in all trials are then analysed as one quasi-stationary data set, using the appropriate rate approximation. (Further details and calibration of the unitary event analysis technique are described in Grün 1996, Grün et al 2001a,b; recent extensions of the approach are discussed in Grün et al 1999, Gütig et al 2001.)

'Unitary events' in motor cortex

In collaboration with Alexa Riehle (CNRS, Marseille, France) we tested the hypothesis that such precise synchronization of individual action potentials among groups of neurons in the monkey motor cortex is involved in dynamically organizing the cortical network during the planning and execution of voluntary movements (Riehle et al 1997).

We found that simultaneously recorded activities of neurons in monkey primary motor cortex indeed exhibited context-dependent, rapid changes in the patterns of coincident action potentials during performance of a delayed-pointing task. Accurate spike synchronization occurred in relation to external events (visual stimuli, hand movements), commonly accompanied by discharge rate modulations, however, without precise time-locking of the spikes to these external events. Accurate spike synchronization also occurred in relation to purely internal events (stimulus expectancy), where firing rate modulations were distinctly absent. These findings indicate that internally generated synchronization of individual spike discharges may subserve the cortical organization of cognitive motor processes. The clear correlation of spike coincidences with stimuli and behavioural events underlines their functional relevance (Riehle et al 1997; see also Fetz 1997).

Taken together, these findings demonstrate the existence of precise (± 1–3 ms) synchronization of individual spike discharges among selected groups of neurons in the motor cortex. This synchronization is associated with distinct phases in the planning and execution of voluntary movements, indicating that it indeed plays a functional role. Moreover, these findings suggest that under behavioural conditions as investigated in this study, the brain uses different strategies in different contextual situations: in order to process a purely cognitive, i.e. an internal and behaviourally relevant event, neurons preferentially synchronize their spike occurrences without changing, at the same time, their firing rates. By contrast, when processing an external, behaviourally relevant event, neurons tend to synchronize their spikes and modulate their firing rates at the same time. Thus, precise synchronization of spike events and modulation of discharge rate may serve different and complementary functions. They act in conjunction at some times, and not others, depending on the behavioural context (Riehle et al 1997).

Conditions for stable propagation of synchronous spiking in cortical networks

In a complementary, model-oriented study we explored the mechanisms underlying these rapid synchronizations of cortical spiking activity. Specifically we focused on the explanation for the excessive occurrences of highly accurate ($\pm$1–3 ms) spike patterns (Abeles et al 1993a,b, Riehle et al 1997, Prut et al 1998), observed in frontal cortex and in motor cortex neurons of awake behaving monkeys.

Synfire chains and pulse packets

On the basis of the characteristic anatomy and physiology of the cortex, Abeles (1982a, 1991) proposed that 'synfire' activity, which propagates in volleys through the sparsely firing cortical neural network, presents a natural explanation for this phenomenon. We have investigated the conditions under which such synchronous volleys of action potentials can propagate reliably through the cortical network (Diesmann et al 1996, 1999, Aertsen et al 1996). Our theoretical approach combined analytical calculations and extensive simulations of single-neuron responses and network dynamics (Diesmann et al 1995, Gewaltig 1999).

Existing measures for the efficacy of synaptic transmission concentrate on two limiting cases: full synchrony and random arrival of spikes. Intermediate cases with a realistic degree of temporal dispersion are hardly addressed. To overcome these restrictions and to quantify the degree of temporal synchrony in propagating volleys of spike activity we introduced the concept of 'pulse packets' (Diesmann et al 1996). A pulse packet is a probabilistic description of the spiking activity of a group of neurons, represented by a pulse density function. This density function is characterized by two parameters: the 'activity', defining the number of spikes in the volley, and the 'width', defining their temporal dispersion. For a single realisation of a pulse packet, the activity is measured by counting the number of spikes in the volley, and its width is measured by the standard deviation of the spike distribution.

Neural transfer function and synchronization dynamics

Adopting this approach, we studied the response behaviour of a model cortical neuron to input activity with varying degrees of synchrony by presenting pulse packets with different choices of the 'activity' and 'width' parameters as stimuli.

From the model neuron we recorded the response (time of first spike), collected in a peri-stimulus time (PST) histogram over many trials. After normalization for the number of trials, the resulting output distribution was again described as a pulse packet, and the associated pulse density, along with the values of the activity and width were determined. The resulting neural transfer function, which describes the input–output relation between incoming and outgoing pulse packets, was visualized in an iterative map. This map yields a compact characterization of the neuron's response to transient input. In contrast to earlier approaches where the neuron's firing probability is measured quasi-statically as a function of DC current, this new transmission function takes full account of the dynamic properties of the input distribution (Aertsen et al 1996).

The temporal evolution of a pulse packet as it travels through the network can be traced by iterating the transfer function. Keeping the width of the chain fixed at a value in the order of 100, the dynamics of the two-dimensional iterated system is characterized by three fix points: two attractors and a saddle point. These fix points partition the state space in two domains, with stable propagation of the synchronous pulse packet in the first and extinction of the synchronous activity in the second. For increasing numbers of neurons per group, the fix points move further apart, increasing the basin of attraction, i.e. the range over which synchronous spiking can survive in the network. By contrast, for too few neurons per group, the fix points disappear, and all trajectories lead to extinction. Synchronous spiking then is no longer a viable option for the network. We found that under physiological conditions, pools of 100 neurons can easily sustain stable synchronous transmission through the network (Diesmann et al 1999).

This state space portrait describes the evolution of synchronous activity 'in the mean', i.e. by subsequent values of the expectation of the pulse packet parameters across trials with different background activity realizations. On the basis of network simulations we could confirm that the results of such analysis in the mean also hold for single-trial realizations (Gewaltig et al 2000, 2001). Around each point of a trajectory, these realizations form a distribution with a width determined by the pulse packet parameters, the group size and inter-group connectivity. This width becomes more important near the separatrix, due to the increased probability — even for trajectories which are stable in the mean — that individual realizations leave the basin of attraction (and vice versa). Thus, it is possible to assess the survival probability at each point in the state space, by computing which fraction of the trajectories crossing a small area around that point reaches the attractor. We found that there is a wide range of stimulus parameters for which the pulse packet is likely to evolve towards the attractor. If the pulse packet is moved away from the fix point, it is able to re-synchronize and to re-gain activity. Important aspects of these synchronization dynamics could be dissected and understood with the help of a continuous, probabilistic description of propagating synfire activity: the 'pulse-density model' (Gewaltig et al 1997, Gewaltig 1999).

Synchronization dynamics in recurrent networks

We also studied the spatiotemporal dynamics of spiking activity in cortical network models with recurrent synaptic architecture (Rotter & Aertsen 2000). The dynamics in such networks provide clues to the interplay that may result from the simultaneous activity of many pulse packets travelling through the cortical network.

Spatiotemporal patterns of precisely timed spikes

We used a network model, which is based on interacting stochastic point processes (Rotter 1994, 1996). Such systems can be formally described in terms of a Markov process, the dynamic state of which at a given point in time is the spatiotemporal pattern of previously generated spikes. The transition probabilities specify how the pattern gradually evolves in time. A generalized type of integrate-and-fire dynamics thereby emerges as a mathematical consequence of the assumption that neurons communicate by action potentials. Assuming the existence of infinitesimal spike probabilities, which is in fact a very mild condition for physical systems, the corresponding dynamic equations could be completely solved.

The solutions for special cases have been used to identify some important model parameters from electrophysiological recordings of real neurons. A simple parametric characterization of single neuron function is in fact achieved by fitting the model to the discharge behaviour of various types of cortical pyramidal cells. Some fundamental properties of recurrent cortex-like networks assembled from such neurons can be readily predicted, most notably their ability to maintain stable low rates of activity without the help of inhibitory neurons (Rotter & Aertsen 1997). Furthermore, computer simulations of random-topology, but otherwise realistic cortical networks indicate that high precision spatiotemporal patterns, embedded in periods of enhanced cooperative group activity, may play a role in coding and computation in such networks. This is true, even if neither the anatomy of the network nor the physiology of its neurons are in any sense specifically designed for that purpose.

Plasticity of precise time structure

Plasticity of the temporal structure of patterns of precisely timed spikes is achieved by introducing Hebb-like synaptic plasticity into the network. The phenomena observed in a number of experiments concerning the influence of local synaptic modification on the spatiotemporal dynamics in recurrent networks allow a number of conclusions (Rotter & Aertsen 1995, Rotter 1996). Learning rules can be formulated which only use local information, without the necessity for explicit renormalization of total synaptic transmission (cf. Song et al 2000, Rubin et al 2001). Evidence for temporally asymmetric plasticity, very much in line with such learning rules, has recently come from electrophysiological studies (Markram et al 1997, Bi & Poo 1998). Using such rules, rapid convergence of synaptic strengths can be achieved, while stable global activity is maintained. Convergence can be extremely fast, within a few presynaptic action potentials. The reason is that the pre-existing (random) patterns of activity are 're-used' or only slightly modified until the correlation structure of the stimulus input is matched. Learning affects only the microscopic time scale, i.e. there is plasticity of time structure in the millisecond range. In fact, the Hebbian time window defining temporal coherence is determined both by the dynamics of after-hyperpolarization in the post-synaptic neuron and by the low-pass properties of the synapse. Modification of a synapse can be enabled and disabled by controlling the rate of the presynaptic neuron. Thereby, a more global strategy of supervised learning is achieved by letting pools of dedicated instructor neurons control firing rates within the network, depending on some reward condition. The learning of input–output associations may take place in terms of a stochastic exploration of error gradients. Again, this amounts to a completely local processing of global information.

Conclusions and outlook

Assuming realistic values for the anatomical and physiological parameters, our model work predicts that the cortical network is able to sustain stable propagation of synchronous spike volleys consisting of spikes from groups of about 100 neurons, interconnected in feedforward fashion, with a temporal precision of about 1 ms. We are currently investigating to what extent the cortical architecture supports the existence of such structures, and how they are spatially embedded in the cortical network (Hehl et al 2001).

Evidence from recent computer simulations suggests that the observed synchronization dynamics are strongly influenced by the activity climate in the surrounding network. In particular, the robustness and propagation velocity of the synchronous spike volleys exhibit a non-monotonic dependence on the level (Diesmann et al 2000) and temporal structure (Mohns et al 1999) of the background activity. With increasing membrane potential fluctuations, the basin of attraction first increases and then decreases again (see also Boven & Aertsen 1990, Aertsen et al 1994), a phenomenon reminiscent of stochastic resonance (Collins et al 1996).

These results have interesting consequences in view of recent findings regarding the relation between ongoing network activity and the variability of evoked responses, both in cortical activity and in behavioural responses (Arieli et al 1996a,b, Azouz & Gray 1999).

Our findings on the synchronization dynamics in recurrent networks indicate that the degree of irregularity of neuronal spike trains is primarily a reflection of the network dynamics. Spatiotemporal patterns of precisely timed spikes are a consequence of these network dynamics. The introduction of Hebb-like synaptic learning rules (cf. Song et al 2000, Rubin et al 2001, Gütig et al 2001) induces a plasticity of the precise spike patterns. Possible scenarios for the functional relevance of such precisely timed spike patterns and their plasticity are the subject of current investigation.

Acknowledgements

This report summarizes results from a number of ongoing collaborations, and extends an earlier review by two of the authors (Rotter & Aertsen 2000). Contributions by and stimulating discussions with Moshe Abeles, George Gerstein, Alexa Riehle and Eilon Vaadia are gratefully acknowledged. Partial funding was received from the Deutsche Forschungsgemeinschaft (DFG), the Human Frontier Science Program (HFSP) and the German-Israeli Foundation for Research and Development (GIF).

References

Abeles M 1982a Local cortical circuits. Springer-Verlag, Berlin
Abeles M 1982b The role of the cortical neuron: integrator or coincidence detector? Isr J Med Sci 18:83–92
Abeles M 1991 Corticonics. Cambridge University Press, Cambridge
Abeles M, Bergman H, Margalit E, Vaadia E 1993a Spatiotemporal firing patterns in the frontal cortex of behaving monkeys. J Neurophysiol 70:1629–1643
Abeles M, Prut Y, Bergman H, Vaadia E, Aertsen A 1993b Integration, synchronicity and periodicity. In: Aertsen A (ed) Brain theory: Spatio-temporal aspects of brain function. Elsevier Science Publishers, Amsterdam, p 149–181
Aertsen A, Arndt M 1993 Response synchronization in the visual cortex. Curr Opin Neurobiol 3:586–594
Aertsen A, Vaadia E, Abeles M et al 1991 Neural interactions in the frontal cortex of a behaving monkey: signs of dependence on stimulus context and behavioral state. J Hirnforsch 32: 735–743
Aertsen A, Erb M, Palm G 1994 Dynamics of functional coupling in the cerebral cortex: an attempt at a model-based interpretation. Physica D 75:103–128
Aertsen A, Diesmann M, Gewaltig MO 1996 Propagation of synchronous spiking activity in feedforward neural networks. J Physiol Paris 90:243–247
Ahissar M, Ahissar E, Bergman H, Vaadia E 1992, Encoding of sound-source location and movement: activity of single neurons and interactions between adjacent neurons in the monkey auditory cortex. J Neurophysiol 67:203–215
Arieli A, Sterkin A, Grinvald A, Aertsen A 1996a Dynamics of ongoing activity: explanation of the large variability in evoked cortical responses. Science 273:1868–1871

Arieli A, Donchin O, Aertsen A et al 1996b The impact of ongoing cortical activity on evoked potentials and behavioral responses in the awake behaving monkey. Soc Neurosci Abstr 22:2022

Azouz R, Gray CM 1999 Cellular mechanisms contributing to response variability of cortical neurons *in vivo*. J Neurosci 19:2209–2223

Barlow HB 1972 Single units and sensation: a neuron doctrine for perceptual psychology? Perception 1:371–394

Barlow HB 1992 Single cells versus neuronal assemblies. In: Aertsen A, Braitenberg VB (eds) Information processing in the cortex. Springer-Verlag, Berlin, p 169–174

Bi GQ, Poo MM 1998 Synaptic modifications in cultured hippocampal neurons: dependence on spike timing, synaptic strength, and postsynaptic cell type. J Neurosci 18:10464–10472

Boven KH, Aertsen A 1990 Dynamics of activity in neuronal networks give rise to fast modulations of functional connectivity. In: Eckmiller R, Hartmann G, Hauske G (eds) Parallel processing in neural systems and computers. Elsevier Science Publishers, Amsterdam, p 53–56

Collins JJ, Chow CC, Capela AC, Imhoff TT 1996 Aperiodic stochastic resonance. Phys Rev E 54:5575–5584

de Charms RC, Merzenich MM 1995 Primary cortical representation by the coordination of action potential timing. Nature 381:610–613

Diesmann M, Gewaltig MO, Aertsen A 1995 SYNOD: An environment for neural systems simulations — language interface and tutorial. Technical Report GC-AA/95–3. The Weizmann Institute of Science, Rehovot, Israel (see also: http//www.synod.uni-freiburg.de)

Diesmann M, Gewaltig MO, Aertsen A 1996 Characterization of synfire activity by propagating "pulse packets". In: Bower J (ed) Computational Neuroscience–Trends in Research 1995. Academic Press, San Diego, p 59–64

Diesmann M, Gewaltig MO, Aertsen A 1999 Stable propagation of synchronous spiking in cortical neural networks. Nature 402:529–533

Diesmann M, Gewaltig MO, Aertsen A 2000 Analysis of spike synchronization in feed-forward cortical neural networks. Soc Neurosci Abstr 26:2201

Eckhorn R, Bauer R, Jordan W et al 1988 Coherent oscillations: a mechanism of feature linking in the visual cortex? Multiple electrode and correlation analysis in the cat. Biol Cybern 60: 121–130

Eggermont JJ 1994 Neural interaction in cat primary auditory cortex II. Effects of sound stimulation. J Neurophysiol 71:246–270

Engel AK, König P, Kreiter AK, Schillen TB, Singer W 1992 Temporal coding in the visual cortex: new vistas on integration in the nervous system. Trends Neurosci 15: 218–226

Fetz EE 1997 Temporal coding in neural populations? Science 278:1901–1902

Georgopoulos AP, Taira M, Lukashin A 1993 Cognitive neurophysiology of the motor cortex. Science 260:47–52

Gerstein GL, Bedenbaugh P, Aertsen A 1989 Neuronal assemblies. IEEE (Inst Electr Electron Eng) Trans Biomed Eng 36:4–14

Gewaltig MO 1999 Evolution of synchronous spike volleys in cortical networks — Network simulations and continuous probabilistic methods. Shaker Verlag, Aachen

Gewaltig MO, Rotter S, Diesmann M, Aertsen A 1997 Cortical synfire activity — Stability in an analytical model. In: Elsner N, Wässle H (eds) From membrane to mind: proceedings of the 25th Göttingen neurobiology meeting 1997. Thieme Verlag, Stuttgart, p 620

Gewaltig MO, Diesmann M, Aertsen A 2000 Cortical synfire activity: Configuration space and survival probability. Neurocomputing, in press

Gewaltig MO, Diesmann M, Aertsen A 2001 Propagation of cortical synfire activity — survival probability in single trials and stability in the mean. Neural Netw, in press

Gray CM, König P, Engel AK, Singer W 1989 Oscillatory responses in cat visual cortex exhibit inter-columnar synchronization which reflects global stimulus patterns. Nature 338:334–337

Grün S 1996 Unitary joint-events in multiple neuron spiking activity — Detection, significance, and interpretation. Verlag Harri Deutsch, Frankfurt

Grün S, Aertsen A, Abeles M, Gerstein G, Palm G 1994 Behavior-related neuron group activity in the cortex. In: Proc 17th Ann Meeting Europ Neurosci Assoc. Oxford Univ Press, Oxford, p 11

Grün S, Diesmann M, Grammont F, Riehle A, Aertsen A 1999 Detecting unitary events without discretization of time. J Neurosci Meth 94:67–79

Grün S, Diesmann M, Aertsen A 2001a 'Unitary events' in multiple single-neuron spiking activity I: detection and significance. Neural Comput

Grün S, Diesmann M, Aertsen A 2001b 'Unitary events' in multiple single-neuron spiking activity II: non-stationary data. Neural Comput, in press

Gütig R, Rotter A, Aertsen A 2001 Statistical significance of coincident spikes: count-based versus rate-based statistics. Neural Comput, in press

Gütig R, Aharonov-Barki R, Rotter A, Aertsen A, Sompolinsky H 2001 Generalized synaptic updating in temporally asymmetric Hebbian learning. In: Göttingen Neurobiology Report 2001, Proc 28th Göttingen Neurobiology Conf, in press

Hebb D 1949 The organization of behavior. A neuropsychological theory. Wiley, New York

Hehl U, Hellwig B, Rotter S, Diesmann M, Aertsen A 2001 Anatomical constraints on stable propagation of synchronous spiking in cortical networks. In: Göttingen Neurobiology Report 2001, Proc 28th Göttingen Neurobiology Conf, in press

Hubel DH, Wiesel TN 1977 Farrier lecture. Functional architecture of macaque monkey visual cortex. Proc R Soc Lond B Biol Sci 198:1–59

Johannesma P, Aertsen A, van den Boogaard H, Eggermont J, Epping W 1986 From synchrony to harmony: Ideas on the function of neural assemblies and on the interpretation of neural synchrony. In: Palm G, Aertsen A (eds) Brain Theory. Springer, Berlin, p 25–47

Mainen ZF, Sejnowski TJ 1995 Reliability of spike timing in neocortical neurons. Science 268:1503–1506

Markram H, Lübke J, Frotscher M, Sakmann B 1997 Regulation of synaptic efficacy by coincidence of postsynaptic APs and EPSPs. Science 275:213–215

Mohns M, Diesmann M, Grün S, Aertsen A 1999 Interaction of synchronous input activity and subthreshold oscillations of membrane potential. In: Elsner N, Eysel U (eds) From molecular neurobiology to clinical neuroscience. Thieme Verlag, Stuttgart, p 100

Murthy VN, Fetz EE 1992 Coherent 25- to 35-Hz oscillations in the sensorimotor cortex of awake behaving monkeys. Proc Natl Acad Sci USA 89:5670–5674

Newsome WT, Britten KH, Movshon JA 1989 Neuronal correlates of a perceptual decision. Nature 341:52–54

Nicolelis MA, Baccala LA, Lin RC, Chapin JK 1995 Sensorimotor encoding by synchronous neural ensemble activity at multiple levels of the somatosensory system. Science 268:1353–1358

Nowak LG, Sanchez-Vives MV, McCormick DA 1997 Influence of low and high frequency inputs on spike timing in visual cortical neurons. Cereb Cortex 7:487–501

Palm G 1990 Cell assemblies as a guideline for brain research. Conc Neurosci 1:133–147

Perkel DH, Bullock TH 1968 Neural Coding. Neurosci Res Program Bull, Vol. 6, Nr. 3. MIT, Brookline, MA

Prut Y, Vaadia E, Bergman H, Haalman I, Slovin H, Abeles M 1998 Spatiotemporal structure of cortical activity: properties and behavioral relevance. J Neurophysiol 79:2857–2874

Riehle A, Grün S, Diesmann M, Aertsen A 1997 Spike synchronization and rate modulation differentially involved in motor cortical function. Science 278:1950–1953

Roelfsema PR, Engel AK, König P, Singer W 1996 The role of neuronal synchronization in response selection: A biologically plausible theory of structured representations in the visual cortex. J Cogn Neurosci 8:603–625

Rotter S 1994 Wechselwirkende stochastische Punktprozesse als Modell für neuronale Aktivität im Neocortex der Säugetiere. Harri Deutsch, Frankfurt

Rotter S 1996 Biophysical aspects of cortical networks. In: Torre V, Conti F (eds) Neurobiology: ionic channels, neurons, and the brain. Plenum Press, New York, p 355–369

Rotter S, Aertsen A 1995 Hebb's rule in networks of spiking neurons. In: Elsner N, Menzel R (eds) Learning and memory. Thieme Verlag, Stuttgart, p 69

Rotter S, Aertsen A 1997 Current balance and neuronal assemblies. In: Elsner N, Wässle H (eds) From membrane to mind: proceedings of the 25th Göttingen neurobiology meeting 1997. Thieme Verlag, Stuttgart, p 622

Rotter S, Aertsen A 2000 Cortical dynamics — Experiments and models. In: Lehnertz K, Arnhold J, Grassberger P, Elger CE (eds) Chaos in brain? World Scientific Publishing, Singapore, p 3–12

Rubin J, Lee D, Sompolinsky H 2001 Equilibrium properties of temporally asymmetric Hebbian plasticity. Phys Rev Lett 86:364–367

Sakurai Y 1996 Population coding by cell assemblies — What it really is in the brain. Neurosci Res 26:1–16

Sanes JN, Donoghue JP 1993 Oscillations in local field potentials of the primate motor cortex during voluntary movement. Proc Natl Acad Sci USA 90:4470–4474

Singer W 1993 Synchronization of cortical activity and its putative role in information processing and learning. Annu Rev Physiol 55:349–374

Singer W, Gray CM 1995 Visual feature integration and the temporal correlation hypothesis. Annu Rev Neurosci 18:555–586

Softky WR, Koch C 1993 The highly irregular firing of cortical cells is inconsistent with temporal integration of random EPSPs. J Neurosci 13:334–350

Song S, Miller KD, Abbott LF 2000 Competitive Hebbian learning through spike-timing-dependent synaptic plasticity. Nat Neurosci 3:919–926

Stevens C, Zador AM 1998 Input synchrony and the irregular firing of cortical neurons. Nat Neurosci 1:210–217

Vaadia E, Haalman I, Abeles M et al 1995 Dynamics of neuronal interactions in monkey cortex in relation to behavioral events. Nature 373:515–518

Volgushev M, Christiakova M, Singer W 1998 Modification of discharge patterns of neocortical neurons by induced oscillations of the membrane potential. Neuroscience 83:15–25

von der Malsburg C 1981 The correlation theory of brain function. Internal Report. MPI Biophys Chem, Göttingen, p 81–82

DISCUSSION

Sejnowski: There is a problem that has to do with the probability of transmission at synapses. Several groups have now used various techniques to look at the reliability of transmission at a single synapse between, for example, two pyramidal cells. It varies. The peak of the distribution is at one tenth: every 10 times that you stimulate the axon, on average you only get the release of a single vesicle on one of those trials. That is a typical synapse. There are some that have a probability of a third or a half, and there are a few that are silent. How does this degree of unreliability at the synapse fit into a model like yours that requires recovery of precision at every stage.

Aertsen: The actual values for the synaptic strengths used in our model were taken from the experimental literature. As these numbers are based on spike-triggered averaging, they of course represent an average picture. As I have shown, the stability is very much governed by the size of the neuron groups in the network. You can compensate for lower synaptic strength by up-scaling this group size. Essentially, it is the product of the two that determines what arrives at the next stage. So, if you bring the synaptic connectivity down, you will need more neurons per group. If, by contrast, you manage to increase the strength of the synaptic connections — through learning or some other means — this will bring the necessary group size down. Another issue is how this scaling interacts with background activity. In additional simulations (Diesmann et al 2000) we found that if you consider the level of background activity, this introduces a third axis, in addition to the two I showed here. As a result, the phase portrait is re-shaped in a rather complex way, because it depends in a non-monotonic way on this third dimension. Yet, there are interesting trade-offs that can be made between the level of background activity and the numbers for the necessary group size and synaptic connectivity.

Berridge: When considering Terry Sejnowski's comment about failures in synaptic transmission, it is reasonable to ask whether there are any data on how many synapses are formed between interacting neurons. Perhaps you get around the failure rate by having more synapses.

Aertsen: There are numbers on this from various sources. Braitenberg was one of the first who looked into this (reviewed in Braitenberg & Schüz 1991), later several others also studied it. The number of synapses between any two neurons in the neo-cortex depends strongly on the distance between the two cells. If they are very close, there is a high probability that they will have multiple (up to 10) synapses between them; if they are further apart ($100\,\mu$m or more), this probability goes down rapidly (e.g. Hellwig 2000). So, neurons that are some $500\,\mu$m apart will typically have at most one synapse between them.

Berridge: Then this probability of failure really matters.

Aertsen: Yes. For a story like this to hold under such circumstances, by necessity we need to increase the size of the assembly. Also, it imposes interesting constraints on the amount of cortical space such an assembly can live in (Hehl et al 2001). I would like to point out that if this doesn't work, nothing does. This is the only viable type of activity in such networks.

Iyengar: I am still thinking what your boundary conditions mean. To achieve that, one has to increase reliability at each synapse, so there is no potentiation but the synapse becomes reliable enough that all of them work, and if this is not achieved in a few cases it fails. If you go back and record at single synapses will they become more reliable?

Sejnowski: That has been done. It is much easier to potentiate a low probability synapse than a high probability synapse. Conversely, it is much easier to depress a high probability synapse. I think there is a close relationship between the two. One idea of LTP is that you are just converting a synapse from a low probability state to a high probability state. This can be deceiving. The beauty of having a contact of a low probability is that you can recruit it if you need it, and you can reorganize your network. There is yet another degree of complexity that underlies synapses that has to do with short-term dynamics. For example, if you stimulate a synapse at high frequency, some synapses will depress, that is each subsequent signal will produce a smaller output, and there are some synapses where the probability of release will go up.

Iyengar: This scares me, because then I wonder how is it that these biochemical events in each of these get coordinated to produce these results.

Aertsen: I agree that the combinatorial complexity increases with each new axis that you open up. On the other hand, part of the good news is that this sort of construction creates robustness.

Sejnowski: There may be a principle for self-repair of a network with many unreliable components, which collectively produces a reliable state.

Iyengar: So you pre-select biochemically for those that are working, and when you reach a critical number the system becomes reliable.

Eichele: However, there are organisms that have very few neurons, yet they still work.

Aertsen: This isn't a theory for all brains of all animals. It is just a theory for the neo-cortex of the mammalian brain. Moreover, it critically depends on the spike rates in the network: it works nicely for low to moderate rates (typical for cortex), but at high spike rates, this theory breaks down.

Sejnowski: Even in humans there are synapses that are highly reliable, such as the neuromuscular junction, which releases so many vesicles that a contraction is bound to occur, regardless of the fluctuation. Where reliability is called for, nature usually achieves this with an anatomical specialization. This is not found in the cortex, except in a few specialized places such as the mossy fibre terminals in CA3.

Laughlin: I would put a slightly different gloss on it. We found that the single synapse, which is just a small synapse, 0.5×0.1 μm, was transmitting 55 bits per second. It is achieving a good transmission rate without any failure. It is not just a question of using large numbers of synapses or big synapses. You can engineer small synapses to be reliable or unreliable, presumably by adjusting vesicle release mechanisms.

Sejnowski: The distinction there is that it is a graded or drip synapse, working over a wide range of potentials, whereas in the cortex it is an all-or-none event.

Laughlin: I would say that these synapses have been deliberately engineered to be unreliable and to have their probability of release depend on other events.

References

Braitenberg V, Schüz A 1991 Anatomy of the cortex. Springer, Berlin

Diesmann M, Gewaltig M-O, Aertsen A 2000 Analysis of spike synchronization in feed-forward cortical neural networks. Soc Neurosci Abstr 26:2201

Hehl U, Hellwig B, Rotter S, Diesmann M, Aertsen 2001 Anatomical constraints on stable propagation of synchronous spiking in cortical networks. In: Göttingen Neurobiology Report 2001. Thieme, Stuttgart, in press

Hellwig B 2000 A quantitative analysis of the local connectivity between pyramidal neurons in layers 2/3 of the rat visual cortex. Biol Cybernetics 82:111–121

Predictive learning of temporal sequences in recurrent neocortical circuits

Rajesh P. N. Rao[*][1] and Terrence J. Sejnowski[†][‡][2]

*Sloan Center for Theoretical Neurobiology, The Salk Institute for Biological Studies, La Jolla, CA 92037, †Howard Hughes Medical Institute, The Salk Institute for Biological Studies, La Jolla, CA 92037 and ‡Department of Biology, University of California at San Diego, La Jolla, CA 92037, USA

Abstract. When a spike is initiated near the soma of a cortical pyramidal neuron, it may back-propagate up dendrites toward distal synapses, where strong depolarization can trigger spike-timing dependent Hebbian plasticity at recently activated synapses. We show that (a) these mechanisms can implement a temporal-difference algorithm for sequence learning, and (b) a population of recurrently connected neurons with this form of synaptic plasticity can learn to predict spatiotemporal input patterns. Using biophysical simulations, we demonstrate that a network of cortical neurons can develop direction selectivity similar to that observed in complex cells in alert monkey visual cortex as a consequence of learning to predict moving stimuli.

2001 Complexity in biological information processing. Wiley, Chichester (Novartis Foundation Symposium 239) p 208–233

Neocortical circuits are dominated by massive excitatory feedback: more than 80% of the synapses made by excitatory cortical neurons are onto other excitatory cortical neurons (Douglas et al 1995, Braitenberg & Schüz 1991). Why is there such massive recurrent excitation in the neocortex and what is its role in cortical computation? Previous modelling studies have suggested a role for excitatory feedback in amplifying feedforward inputs (Douglas et al 1995, Suarez et al 1995, Mineiro & Zipser 1998, Ben-Yishai et al 1995, Somers et al 1995, Chance et al 1999). Recently, it has been shown that recurrent excitatory connections between cortical neurons are modified according to a spike-timing dependent Hebbian learning rule: synapses that are activated slightly before the cell fires are

---

[1]Present address: Department of Computer Science & Engineering, University of Washington, 114 Sieg Hall, Seattle, WA 98195-2350, USA.
[2]This paper was presented at the Symposium by Terrence J. Sejnowski, to whom correspondence should be addressed.

strengthened whereas those that are activated slightly after are weakened (Markram et al 97) (see also Levy & Steward 1983, Zhang et al 1998, Bi & Poo 1998, Abbott & Blum 1996, Gerstner et al 1996, Senn 1997). Information regarding the postsynaptic activity of the cell is conveyed back to the dendritic locations of synapses by backpropagating action potentials from the soma (Stuart & Sakmann 1994).

Because these recurrent feedback connections can adapt in a temporally specific manner, they may subserve a more general function than amplification, such as the prediction and generation of temporal sequences (Abbott & Blum 1996, Minai & Levy 1993, Montague & Sejnowski 1994, Schultz et al 1997, Softky 1996, Koch 1999, Rao & Ballard 1997). The observation that recurrence can generate sequences has its roots in dynamical systems theory (Scheinerman 1995) and forms the basis of numerous engineering (Kalman 1960) and neural network (Minai & Levy 1993, Rao & Ballard 1997, Jordan 1986, Elman 1990) models for predicting and tracking input sequences. Consider the network of excitatory neurons shown in Fig. 1A. By appropriately learning its recurrent connections, the network can generate sequences of outputs in anticipation of its inputs as depicted in Fig. 1B. The initial activation of a subset of input neurons causes the corresponding set of excitatory neurons to be activated, which in turn activate a different set of excitatory neurons and so on, such that each set of active neurons at a given time step represents the anticipated input at that time step (active neurons are represented as shaded circles in Fig. 1B). The predicted outputs occur just in time to inhibit the input neurons if the external input is excitatory, or excite them if the external input is inhibitory, thereby implementing a stable negative feedback loop and allowing only the unpredicted part of the input to be conveyed to the prediction neurons. Such a model is consistent with some recent ideas regarding cortico-cortical feedback loops (Rao & Ballard 1997, Mumford 1994), predictive coding (Rao & Ballard 1999, Barlow 1998, Daugman & Downing 1995) and visual receptive field development from natural images (Rao & Ballard 1997, Olshausen & Field 1997). In these models, feedback connections from a higher to a lower order cortical area are posited to carry predictions of lower level neural activity, while the feedforward connections are assumed to convey the residual errors in prediction. These errors are used to correct the neural representation at the higher level before generating a subsequent prediction (for example, see Rao & Ballard 1997). Note that for clarity, Fig. 1B shows two different sets of excitatory neurons firing at the two successive time steps, but the model allows arbitrary overlapping subsets of neurons to fire in order to represent temporal sequences with possible overlapping inputs, resulting in sustained firing in some neurons and transient firing in others due to learned recurrent connections.

In this study, we have modelled spike-timing dependent Hebbian synaptic plasticity as a form of 'temporal-difference' learning (Montague & Sejnowski

A

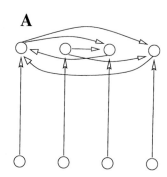

B
Excitatory
Neurons

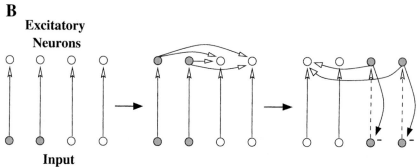

Input

FIG. 1.   Prediction using recurrent excitation. (A) An example of a model network of recurrently connected excitatory neurons receiving inputs from a set of input neurons (bottom row). (B) The activation of a subset of input neurons (shaded circles) causes a subset of excitatory neurons to fire which in turn cause a different subset of excitatory neurons to fire due to recurrent excitatory connections. If these recurrent connections are appropriately learned, the second subset of neurons will fire slightly before the expected activation of their corresponding input neurons, allowing inhibition of the inputs and forming a stable negative feedback loop. For clarity, the example shows two different sets of excitatory neurons firing at the two successive time steps, but the learning algorithm allows arbitrary overlapping subsets of neurons to fire in order to represent sequences with possible overlapping inputs, resulting in sustained firing in some neurons and transient firing in others due to the learned recurrent connections.

1994, Schultz et al 1997, Sutton 1988). We have simulated recurrent networks of excitatory and inhibitory cortical neurons possessing this form of synaptic plasticity and have investigated the ability of such networks to learn predictive models of input sequences, focusing in particular on moving stimuli. Detailed compartmental models take into account the temporal dynamics of signal processing in dendrites and the relative timing of spikes in neural populations. Both of these properties were found to be essential in explaining the genesis of complex cell-like direction selectivity in model neocortical neurons.

Results

Spike-timing dependent Hebbian plasticity as temporal-difference learning

To accurately predict input sequences, the recurrent excitatory connections between a given set of neurons need to be adjusted such that the appropriate set of neurons are activated at each time step. This can be achieved by using a 'temporal-difference' learning rule (Montague & Sejnowski 1994, Schultz et al 1997, Sutton 1988). In this paradigm of synaptic plasticity, an activated synapse is strengthened or weakened based on whether the difference between two temporally separated predictions is positive or negative. This minimizes the errors in prediction by ensuring that the prediction generated by the neuron after synaptic modification is closer to the desired value than before (see Methods for more details).

In order to ascertain whether spike-timing dependent Hebbian learning in cortical neurons can be interpreted as a form of temporal-difference learning, we used a two-compartment model of a cortical neuron consisting of a dendrite and a soma-axon compartment. The compartmental model was based on a previous study that demonstrated the ability of such a model to reproduce a range of cortical response properties (Mainen & Sejnowski 1996). Figures 2A and 2B illustrate the responses of the model neuron to constant current pulse injection into the soma and random excitatory and inhibitory Poisson-distributed synaptic inputs to the dendrite respectively (see Methods). The presence of voltage-activated sodium channels in the dendrite allowed backpropagation of action potentials from the soma into the dendrite as shown in Fig. 2C.

To study synaptic plasticity in the model, excitatory postsynaptic potentials (EPSPs) were elicited at different time delays with respect to postsynaptic spiking by presynaptic activation of a single excitatory synapse located on the dendrite. Synaptic currents were calculated using a kinetic model of synaptic transmission (Destexhe et al 1997) with model parameters fitted to whole-cell recorded AMPA (α-amino-3-hydroxy-5-methyl-4-isoxazole proprionic acid) currents (see Methods for more details). Other inputs representing background activity were modelled as sub-threshold excitatory and inhibitory Poisson processes with a mean firing rate of 3 Hz. Synaptic plasticity was simulated by incrementing or decrementing the value for maximal synaptic conductance by an amount proportional to the temporal-difference in the postsynaptic membrane potential at time instants $t + \Delta t$ and $t - \Delta t$ for presynaptic activation at time t (see Methods). The delay parameter Δt was set to 5 ms for these simulations; similar results were obtained for other values in the 5–15 ms range. Presynaptic input to the model neuron was paired with postsynaptic spiking by injecting a depolarizing current pulse (10 ms, 200 pA) into the soma. Changes in synaptic efficacy were monitored by applying a test stimulus before and after pairing, and recording the EPSP evoked by the test stimulus.

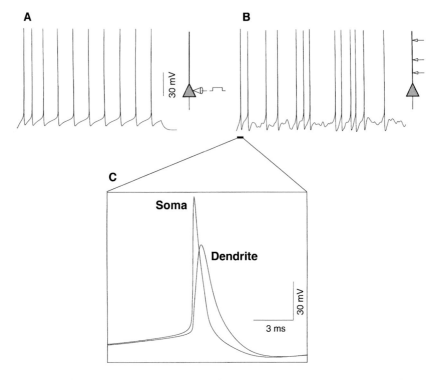

FIG. 2.   Model neuron response properties. (A) Response of a model neuron to a 70 pA current pulse injection into the soma for 900 ms. (B) Response of the same model neuron to Poisson distributed excitatory and inhibitory synaptic inputs at random locations on the dendrite. (C) Example of a backpropagating action potential in the dendrite of the model neuron as compared to the corresponding action potential in the soma (enlarged from the initial portion of the trace in [B]).

Figure 3A shows the results of pairings in which the postsynaptic spike was triggered 5 ms after and 5 ms before the onset of the EPSP, respectively. While the peak EPSP amplitude was increased 58.5% in the former case, it was decreased 49.4% in the latter case, qualitatively similar to experimental observations (Markram et al 1997). As mentioned above, such changes in synaptic efficacy in the model are determined by the temporal-difference in the dendritic membrane potential at time instants $t + \Delta t$ and $t - \Delta t$ for presynaptic activation at time t: the difference is positive when presynaptic activation occurs a few milliseconds before a backpropagating action potential invades the dendrite and negative when it occurs slightly after, causing respectively an increase or decrease in synaptic efficacy. The critical window for synaptic modifications in the model depends on the parameter Δt as well as the shape of the backpropagating action potential. This window of plasticity was examined by

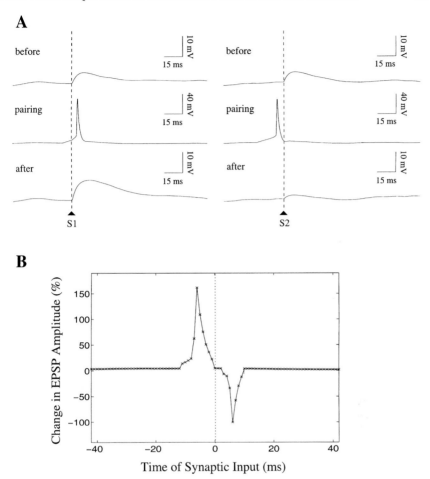

FIG. 3. Synaptic plasticity in a model neocortical neuron. (A) (*Left panel*) The response at the top ('before') is the EPSP evoked in the model neuron due to a presynaptic spike (S1) at an excitatory synapse. Pairing this presynaptic spike with postsynaptic spiking after a 5 ms delay ('pairing') induces long-term potentiation as revealed by an enhancement in the peak of the EPSP evoked by presynaptic simulation alone ('after'). (*Right panel*) If presynaptic stimulation (S2) occurs 5 ms after postsynaptic firing, the synapse is weakened resulting in a decrease in peak EPSP amplitude. (B) Critical window for synaptic plasticity obtained by varying the delay between presynaptic and postsynaptic spiking (negative delays refer to cases where the presynaptic spike occurred before the postsynaptic spike).

varying the time interval between presynaptic stimulation and postsynaptic spiking (with $\Delta t = 5$ ms). As shown in Fig. 3B, changes in synaptic efficacy exhibited a highly asymmetric dependence on spike timing similar to physiological data (Bi & Poo 1998). Potentiation was observed for EPSPs that occurred between 1 and 12 ms before the postsynaptic spike, with maximal potentiation at 6 ms. Maximal depression was observed for EPSPs occurring 6 ms after the peak of the postsynaptic spike and this depression gradually decreased, approaching zero for delays greater than 10 ms. As in rat neocortical neurons (Markram et al 1997), *Xenopus* tectal neurons (Zhang et al 1998), and cultured hippocampal neurons (Bi & Poo 1998), a narrow transition zone (roughly 3 ms in the model) separated the potentiation and depression windows. Note that the exact duration of the potentiation and depression windows in the model can be adapted to match physiological data by appropriately choosing the temporal-difference parameter Δt and/or varying the distribution of active channels in the dendrite the synapse is located on.

Learning to predict using temporal-difference learning

To see how a network of model neurons can learn to predict sequences using the learning mechanism described above, consider the simplest case of two excitatory neurons N1 and N2 connected to each other, receiving inputs from two separate input neurons I1 and I2 (Fig. 4A). Suppose input neuron I1 fires before input neuron I2, causing neuron N1 to fire (Fig. 4B). The spike from N1 results in a sub-threshold EPSP in N2 due to the synapse S2. If input arrives from I2 any time between 1 and 12 ms after this EPSP and the temporal summation of these two EPSPs causes N2 to fire, the synapse S2 will be strengthened. The synapse S1, on the other hand, will be weakened because the EPSP due to N2 arrives a few milliseconds after N1 has fired. Thus, on a subsequent trial, when input I1 causes neuron N1 to fire, it in turn causes N2 to fire several milliseconds *before* input I2 occurs due to the potentiation of the recurrent synapse S2 in previous trial(s) (Fig. 4C). Input neuron I2 can thus be inhibited by the predictive feedback from N2 just before the occurrence of imminent input activity (marked by an asterisk in Fig. 4C). This inhibition prevents input I2 from further exciting N2. Similarly, a positive feedback loop between neurons N1 and N2 is avoided because the synapse S1 was weakened in previous trial(s) (see arrows in Figs 4B and 4C). Figure 4D depicts the process of potentiation and depression of the two synapses as a function of the number of exposures to the I1–I2 input sequence. The decrease in latency of the predictive spike elicited in N2 with respect to the timing of input I2 is shown in Fig. 4E. Notice that before learning, the spike occurs 3.2 ms after the occurrence of the input whereas after learning, it occurs 7.7 ms before the input. This simple example helps to illustrate how subsets of neurons may learn to selectively trigger other subsets of neurons in anticipation of future inputs while maintaining stability in the recurrent network.

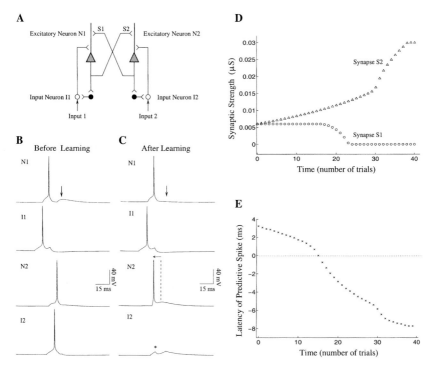

FIG. 4. Learning to predict using spike-timing dependent Hebbian plasticity. (A) A simple network of two model neurons N1 and N2 recurrently connected via excitatory synapses S1 and S2. Sensory inputs are relayed to the two model neurons by input neurons I1 and I2. Feedback from N1 and N2 inhibit the input neurons via inhibitory interneurons (darkened circles). (B) Activity in the network elicited by the input sequence I1 followed by I2. Notice that N2 fires after its input neuron I2 has fired. (C) Activity in the network elicited by the same input sequence after 40 trials of learning. Notice that due to the strengthening of synapse S2, neuron N2 now fires several milliseconds before the time of expected input from I2 (dashed line), allowing it to inhibit I2 (asterisk). On the other hand, synapse S1 has been weakened, thereby preventing re-excitation of N1 (downward arrows show the corresponding decrease in EPSP). (D) Potentiation and depression of synapses S1 and S2 respectively during the course of learning. Synaptic strength was defined as maximal synaptic conductance in the kinetic model of synaptic transmission (see Methods). (E) Latency of the predictive spike in neuron N2 during the course of learning measured with respect to the time of input spike in I2 (dotted line). Note that the latency is initially positive (N2 fires after I2) but later becomes negative, reaching a value of up to 7.7 ms before input I2 as a consequence of learning.

Direction selectivity from predictive sequence learning

To facilitate comparison with published neurophysiological data, we have focused specifically on the problem of predicting moving visual stimuli. Previous modelling studies have suggested that recurrent excitation may play a crucial role in generating direction selectivity in cortical neurons by amplifying their weak

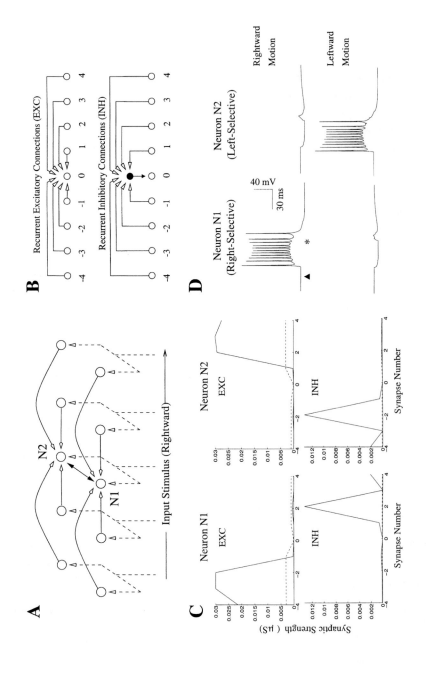

feedforward inputs (Douglas et al 1995, Suarez et al 1995, Mineiro & Zipser 1998). Our simulations suggest that a network of recurrently connected neurons can develop direction selectivity as a consequence of learning to predict moving stimuli. We used a network of recurrently connected excitatory neurons as shown in Fig. 5A receiving retinotopic sensory input consisting of moving pulses of excitation (8 ms pulse of excitation at each neuron) in the rightward and leftward directions. The task of the network was to predict the sensory input by learning appropriate recurrent connections such that a given neuron in the network can fire a few milliseconds before the arrival of its input pulse of excitation. The network was comprised of two parallel chains of neurons with mutual inhibition (dark arrows) between corresponding pairs of neurons along the two chains. The network was initialized such that within a chain, a given excitatory neuron received both excitation and inhibition from its predecessors and successors. This is shown in Fig. 5B for a neuron labelled '0'. Inhibition at a given neuron was mediated by an inhibitory interneuron (dark circle) which received excitatory connections from neighbouring excitatory neurons (Fig. 5B, lower panel). The interneuron received the same input pulse of excitation as the nearest excitatory neuron. Excitatory and inhibitory synaptic currents were calculated using kinetic models of synaptic transmission based on properties of AMPA and $GABA_A$ (γ-aminobutyric acid A) receptors as determined from whole-cell recordings (see Methods). Maximum conductances for all synapses were initialized to small positive values (dotted lines in Fig. 5C) with a slight asymmetry in the recurrent excitatory connections for breaking symmetry between the two chains. The initial asymmetry elicited a single spike slightly earlier for neurons in one chain than neurons in the other chain for a given motion direction, allowing activity in the other chain to be inhibited.

FIG. 5.   Emergence of direction selectivity in the model. (A) A model network consisting of two chains of recurrently connected neurons receiving retinotopic inputs. A given neuron receives recurrent excitation and recurrent inhibition (white-headed arrows) as well as inhibition (dark-headed arrows) from its counterpart in the other chain. (B) Recurrent connections to a given neuron (labelled '0') arise from 4 preceding and 4 succeeding neurons in its chain. Inhibition at a given neuron is mediated via a GABAergic interneuron (darkened circle). (C) Synaptic strength of recurrent excitatory (EXC) and inhibitory (INH) connections to neurons N1 and N2 before (dotted lines) and after learning (solid lines). Synapses were adapted during 100 trials of exposure to alternating leftward and rightward moving stimuli. (D) Responses of neurons N1 and N2 to rightward and leftward moving stimuli. As a result of learning, neuron N1 has become selective for rightward motion (as have other neurons in the same chain) while neuron N2 has become selective for leftward motion. In the preferred direction, each neuron starts firing several milliseconds before the actual input arrives at its soma (marked by an asterisk) due to recurrent excitation from preceding neurons. The dark triangle represents the start of input stimulation in the network.

To evaluate the consequences of synaptic plasticity, the network of neurons was exposed alternately to leftward and rightward moving stimuli for a total of 100 trials. The excitatory connections (labelled 'EXC' in Fig. 5B) were modified according to the asymmetric Hebbian learning rule in Fig. 3B while the excitatory connections onto the inhibitory interneuron (labelled 'INH') were modified according to an asymmetric anti-Hebbian learning rule that reversed the polarity of the rule in Fig. 3B. In other words, if presynaptic activity occurred before (after) the postsynaptic spike in the interneuron, the excitatory connection to the inhibitory interneuron was weakened (strengthened). Although not yet reported in the neocortex, such a rule for inhibitory interneurons is consistent with the spike-timing dependent anti-Hebbian plasticity observed in inhibitory interneurons in a cerebellum-like structure in weakly electric fish (Bell et al 1997).

The synaptic conductances learned by two neurons (marked N1 and N2 in Fig. 5A) located at corresponding positions in the two chains after 100 trails of exposure to the moving stimuli are shown in Fig. 5C (solid line). Initially, for rightward motion, the slight asymmetry in the initial excitatory connections of neuron N1 allows it to fire slightly earlier than neuron N2 thereby inhibiting neuron N2. Additionally, since the EPSPs from neurons lying on the left of N1 occur before N1 fires, the excitatory synapses from these neurons are strengthened while the excitatory synapses from these same neurons to the inhibitory interneuron are weakened according to the two learning rules mentioned above. On the other hand, the excitatory synapses from neurons lying on the right side of N1 are weakened while inhibitory connections are strengthened since the EPSPs due to these connections occur after N1 has fired. The synapses on neuron N2 and its associated interneuron remain unaltered since there is no postsynaptic firing (due to inhibition by N1) and hence no backpropagating action potentials in the dendrite. Similarly, for leftward motion, neuron N2 inhibits neuron N1 and the synapses associated with N2 are adapted according to the two learning rules. As shown in Fig. 5C, after 100 trials, the excitatory and inhibitory connections to neuron N1 exhibit a marked asymmetry, with excitation originating from neurons on the left and inhibition from neurons on the right. Neuron N2 exhibits the opposite pattern of connectivity.

As expected from the learned pattern of connectivity, neuron N1 was found to be selective for rightward motion while neuron N2 was selective for leftward motion (Fig. 5D). Moreover, when stimulus motion is in the preferred direction, each neuron starts firing a few milliseconds before the time of arrival of the input stimulus at its soma (marked by an asterisk) due to recurrent excitation from preceding neurons. Conversely, motion in the non-preferred direction triggers recurrent inhibition from preceding neurons as well as inhibition from the active neuron in the corresponding position in the other chain. Thus, the learned pattern of connectivity allows the direction-selective neurons comprising the two chains in the network to conjointly code for and predict the moving input stimulus in each direction.

The role of recurrent excitation and inhibition

To ascertain the role of recurrent excitation in the model, we gradually decreased the value of the maximum synaptic conductance between excitatory neurons in the network, starting from 100% of the learned values. For a stimulus moving in the preferred direction, decreasing the amount of recurrent excitation increased the latency of the first spike in a model neuron and decreased the spike count until, with less than 10% of the learned recurrent excitation, the latency equalled the arrival time of the input stimulus and the spike count dropped to 1 (Figs 6A and 6B). These results demonstrate that recurrent excitation plays a crucial role in generating predictive activity in model neurons and enhances direction-selective responses by increasing the spike count in the preferred direction.

To evaluate the role of inhibition in maintaining direction selectivity in the model, we quantified the degree of direction selectivity using the direction index: 1−(number of spikes in non-preferred direction)/(number of spikes in preferred direction). Figures 6C and 6D show the distribution of direction indices with and without inhibition in a network of two chains containing 35 excitatory and 35 inhibitory neurons. In the control case, most of the excitatory neurons and inhibitory interneurons receiving recurrent excitation are highly direction selective. Blocking inhibition significantly reduces direction selectivity in the model neurons but does not completely eliminate it, consistent with some previous physiological observations (Sillito 1975, Nelson et al 1994). The source of this residual direction selectivity in the model in the absence of inhibition can be traced to the asymmetric recurrent excitatory connections in the network which remain unaffected by the blockage of inhibition.

Comparison with awake monkey complex cell responses

Similar to complex cells in primary visual cortex, model neurons are direction selective throughout their receptive field because at each retinotopic location, the corresponding neuron in the chain receives the same pattern of asymmetric excitation and inhibition from its neighbours as any other neuron in the chain. Thus, for a given neuron, motion in any local region of the chain will elicit direction-selective responses due to recurrent connections from that part of the chain. This is consistent with previous modelling studies (Chance et al 1999) suggesting that recurrent connections may be responsible for the spatial-phase invariance of complex cell responses. Assuming a 200 μm separation between excitatory model neurons in each chain and utilizing known values for the

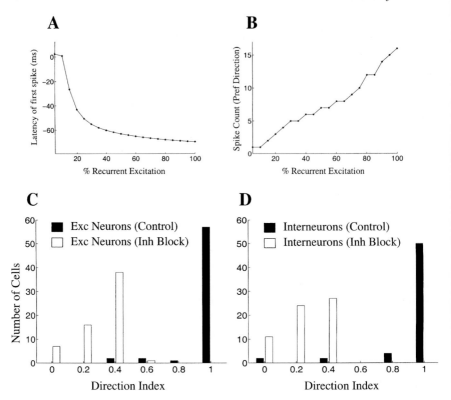

FIG. 6.   The role of recurrent excitation and inhibition. (A) & (B) Latency of the first spike and number of spikes elicited in an excitatory neuron in the preferred direction as a function of the strength of recurrent excitation in a model network (100% corresponds to the learned values of recurrent connection strength). The network comprised of two chains, each containing 35 excitatory neurons and 35 inhibitory interneurons (mutual inhibition between corresponding neurons in the two chains was mediated by a separate set of inhibitory neurons that were not plastic). (C,D) Distribution of direction selectivity in the network for excitatory and inhibitory interneurons respectively with (Control) and without GABAergic inhibition (Inh Block) as measured by the direction index: 1−(Non-Preferred Direction Response)/(Preferred Direction Response).

cortical magnification factor in monkey striate cortex (Tootell et al 1988), one can estimate the preferred stimulus velocity of model neurons to be $3.1°/s$ in the fovea and $27.9°/s$ in the periphery (at an eccentricity of $8°$). Both of these values fall within the range of monkey striate cortical velocity preferences ($1°/s$ to $32°/s$) (van Essen 1985, Livingstone 1998).

The model predicts that the neuroanatomical connections for a direction-selective neuron should exhibit a pattern of asymmetrical excitation and inhibition similar to Fig. 5C. A recent study of direction-selective cells in awake monkey V1 found excitation on the preferred side of the receptive field and inhibition on the null side consistent with the pattern of connections learned by the model (Livingstone 1998). For comparison with this experimental data, spontaneous background activity in the model was generated by incorporating Poisson-distributed random excitatory and inhibitory alpha synapses on the dendrite of each model neuron. Post-stimulus time histograms (PSTHs) and space-time response plots were obtained by flashing optimally oriented bar stimuli at random positions in the cell's activating region. As shown in Fig. 7, there is good qualitative agreement between the response plot for a direction-selective complex cell and that for the model. Both space-time plots show a progressive shortening of response onset time and an increase in response transiency going in the preferred direction; in the model, this is due to recurrent excitation from progressively closer cells on the preferred side. Firing is reduced to below background rates 40–60 ms after stimulus onset in the upper part of the plots; in the model, this is due to recurrent inhibition from cells on the null side. The response transiency and shortening of response time course appears as a slant in the space-time maps, which can be related to the neuron's velocity sensitivity (see Livingstone 1998 for more details).

Discussion

Our results show that a network of recurrently connected neurons endowed with a temporal-difference based asymmetric Hebbian learning mechanism can learn a predictive model of its spatiotemporal inputs. Using a biophysical model of neocortical neurons, we showed that a temporal-difference learning rule for prediction when applied to backpropagating action potentials in dendrites produces asymmetric learning windows similar to those observed in physiological experiments (see Senn 1997, Egelman & Montague 1998) for possible biophysical mechanisms based on N-methyl-D-aspartate (NMDA) receptor activation and voltage-dependent Ca^{2+} channels). When exposed to moving stimuli, neurons in a simulated network with recurrent excitatory and inhibitory connections learned to fire a few milliseconds before the expected arrival of an input stimulus and developed direction selectivity as a consequence of learning. The model predicts that a direction-selective neuron should start responding a few milliseconds before the preferred stimulus arrives at the retinotopic location of the neuron in primary visual cortex. Such predictive neural activity has recently been reported in ganglion cells in the rabbit and salamander retina (Berry et al 1999).

The development of direction selectivity in our model requires a slight initial bias in cortical connectivity (Fig. 5C) which is then sharpened by visual experience of moving stimuli. This is consistent with experimental evidence

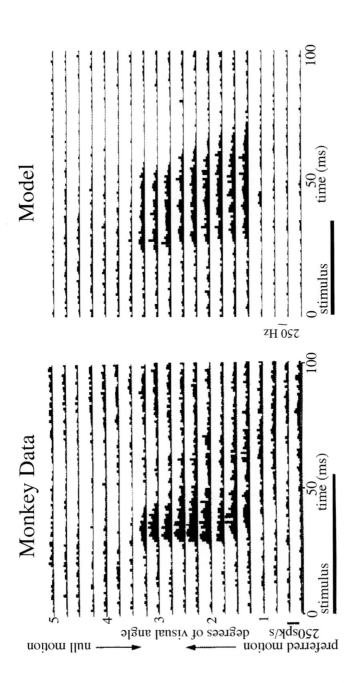

FIG. 7. Comparison of monkey and model space–time response plots. (*Left*) Sequence of PSTHs obtained by flashing optimally oriented bars at 20 positions across the 5°-wide receptive field (RF) of a complex cell in alert monkey V1 (from Livingstone 1998). The cell's preferred direction is from the part of the RF represented at the bottom towards the top. Flash duration = 56 ms; inter-stimulus delay = 100 ms; 75 stimulus presentations. (*Right*) PSTHs obtained from a model neuron after stimulating the chain of neurons at 20 positions to the left and right side of the given neuron. Lower PSTHs represent stimulations on the preferred side while upper PSTHs represent stimulations on the null side.

indicating that (a) some cells in cat visual cortex show some amount of direction selectivity before eye opening (Movshon & van Sluyters 1981) and (b) visual experience during a critical period can profoundly affect the development of direction selectivity (for example, direction selectivity can be abolished by strobe rearing; Humphrey & Saul 1998). Although several models for the development of direction selectivity have been proposed (Feidler et al 1997, Wimbauer et al 1997), the roles of spike timing and asymmetric Hebbian plasticity have not been previously explored. An interesting question currently being investigated is whether the explicit dependence of visual development on spike timing in our model can account for the fact that only low frequencies of stroboscopic illumination (approximately 8 Hz or below) lead to a loss of direction selectivity.

Temporally asymmetric Hebbian learning has previously been suggested as a possible mechanism for sequence learning in the hippocampus (Levy & Steward 1983, Abbott & Blum 1996) and as an explanation for the asymmetric expansion of hippocampal place fields during route learning (Mehta et al 1997). Some of these theories require relatively long temporal windows of synaptic plasticity (on the order of several hundreds of milliseconds) (Abbott & Blum 1996) while others have utilized temporal windows in the sub-millisecond range for coincidence detection (Gerstner et al 1996). Prediction and sequence learning in our model is based on a window of plasticity in the tens of milliseconds range which is roughly consistent with recent physiological observations (Markram et al 1997, Zhang et al 1998, Bi & Poo 1998). Although a fixed learning window (roughly 15 ms of potentiation/depression) was used in the simulations, the temporal extent of this window can be modified by changing the parameter Δt. The temporal-difference model predicts that the shape and width of the asymmetric learning window should be a function of the backpropagating action potentials in the dendrite that the synapse is located on. In the case of hippocampal neurons and cortical neurons, the width of backpropagating action potentials in apical dendrites has been reported to be in the range of 10–25 ms, which is comparable to the size of potentiation/depression windows for synapses located on these dendrites (Bi & Poo 1998, Stuart & Sakmann 1994). Additionally, in order to account for the off regions observed in the receptive fields of cortical direction-selective cells (Livingstone 1998), we included synaptic plasticity of excitatory synapses on inhibitory interneurons by assuming that the sign of the spike-timing dependent Hebbian learning window was inverted from that found on pyramidal neurons. This inversion has been found in excitatory synapses on inhibitory interneurons in a cerebellum-like brain structure in weakly electric fish (Bell et al 1997), but remains a prediction of our model for the cortex.

In vitro experiments involving cortical and hippocampal slices suggest the possibility of short-term plasticity in synaptic connections onto pyramidal neurons (Thomson & Deuchars 1994, Tsodyks & Markram 1997, Abbott et al

1997). The kinetic model of synaptic transmission used in the present study can be extended to include short-term plasticity with the addition of a parameter governing the level of depression caused by each presynaptic action potential (Chance et al 1999, Tsodyks & Markram 1997, Abbott et al 1997). The adaptation of this parameter may allow finer control of postsynaptic firing in the model in addition to the coarse-grained control offered by modifications of maximal synaptic conductance. As suggested by previous studies (Chance et al 1999, Abbott 1997), we expect the addition of synaptic depression in our model to enhance the transient response of model neurons to stimuli such as flashed bars (see Fig. 7) and to broaden the response to drifting stimuli, due to the reduced sensitivity of postsynaptic neurons to mean presynaptic firing rates. In preliminary simulations, the inclusion of short-term plasticity did not significantly alter the development of direction selectivity in recurrent network models as reported here.

The idea that prediction and sequence learning may constitute an important goal of the neocortex has previously been suggested in the context of statistical and information theoretic models of cortical processing (Minai & Levy 1993, Montague & Sejnowski 1994, Mumford 1994, Daugman & Downing 1995, Abbott & Blum 1996, Schultz et al 1997, Rao & Ballard 1997, Barlow 1998, Rao 1999). Our biophysical simulations suggest a possible implementation of such models in cortical circuitry. Several authors have observed the general uniformity in the basic structure of the neocortex across different cortical areas (Hubel & Wiesel 1974, Creutzfeldt 1977, Sejnowski 1986, Douglas et al 1989). Given the universality of the problem of encoding and generating temporal sequences in both sensory and motor domains, the hypothesis of predictive sequence learning in recurrent neocortical circuits may help provide a unifying principle for understanding the general nature of cortical information processing (Creutzfeldt 1977, Sejnowski 1986).

Methods

Temporal-difference learning. The simplest example of a temporal-difference learning rule arises in the problem of predicting a scalar quantity z using a neuron with synaptic weights $w(1), \ldots w(k)$ (represented as a vector $\mathbf{w}$). The neuron receives as presynaptic input the sequence of vectors $\mathbf{x}_1, \ldots \mathbf{x}_m$. The output of the neuron at time t is assumed to be given by: $P_t = \sum_i w(i) x_t(i)$. The goal is to learn a set of synaptic weights such that the prediction P_t is as close as possible to the target z. One method for achieving this goal is to use a temporal-difference (TD[0]) learning rule (Sutton 1988). The weights are changed at time t by an amount given by:

$$\Delta \mathbf{w}_t = \alpha(P_{t+1} - P_t)\mathbf{x}_t \tag{1}$$

where α is a learning rate or gain parameter and the final prediction P_{m+1} is defined to be z. Note that in such a learning paradigm, synaptic plasticity is governed by the temporal difference in postsynaptic activity at time instants $t+1$ and t in conjunction with presynaptic activity $\mathbf{x}_t$ at time t.

Neocortical neuron model. Two-compartment model neocortical neurons consisting of a dendritic compartment and a soma-axon compartment (Mainen & Sejnowski 1996) were implemented using the simulation software *Neuron* (Hines 1993). Four voltage-dependent currents and one Ca^{2+}-dependent current were simulated: fast Na^+, I_{Na}; fast K^+, I_{Kv}; slow non-inactivating K^+, I_{Km}; high voltage-activated Ca^{2+}, I_{Ca}; and Ca^{2+}-dependent K^+ current, I_{KCa} (see Mainen & Sejnowski 1996 for references). Conventional Hodgkin–Huxley-type kinetics were used for all currents (integration time step $=25\,\mu s$, temperature $=37\,°C$). Ionic currents I were calculated using the ohmic equation: $I = \bar{g}A^x B(V-E)$ where $\bar{g}$ is the maximal ionic conductance density, A and B are activation and inactivation variables, respectively (x denotes the order of kinetics; see Mainen & Sejnowski 1996 for further details), and E is the reversal potential for the given ion species ($E_K = -90\,mV$, $E_{Na} = 60\,mV$, $E_{Ca} = 140\,mV$, $E_{leak} = -70\,mV$). The following active conductance densities were used in the dendritic compartment (in $pS/\mu m^2$): $\bar{g}_{Na} = 20$, $\bar{g}_{Ca} = 0.2$, $\bar{g}_{Km} = 0.1$, and $\bar{g}_{KCa} = 3$, with leak conductance $33.3\,\mu S/cm^2$ and specific membrane resistance $30\,k\Omega/cm^2$. The soma–axon compartment contained $\bar{g}_{Na} = 40\,000$ and $\bar{g}_{Kv} = 1400$. For all compartments, the specific membrane capacitance was $0.75\,\mu F/cm^2$. Two key parameters governing the response properties of the model neuron are (Mainen & Sejnowski 1996): the ratio of axo-somatic area to dendritic membrane area (ρ) and the coupling resistance between the two compartments (κ). For the present simulations, we used the values $\rho = 150$ (with an area of $100\,\mu m^2$ for the soma–axon compartment) and a coupling resistance of $\kappa = 8\,M\Omega$. Poisson-distributed synaptic inputs to the dendrite were simulated using alpha function (Koch 1999) shaped current pulse injections (time constant $=5\,ms$) at Poisson intervals with a mean presynaptic firing frequency of $3\,Hz$.

Model of synaptic transmission and plasticity. Synaptic transmission at excitatory (AMPA) and inhibitory (GABA$_A$) synapses was simulated using first order kinetics of the form:

$$\frac{dr}{dt} = \alpha[T](1-r) - \beta r \tag{2}$$

where $r(t)$ denotes the fraction of postsynaptic receptors bound to the neurotransmitter at time t, $[T]$ is the neurotransmitter concentration, and α and β are the forward and backward rates for transmitter binding. Assuming receptor binding directly gates the opening of an associated ion channel, the resulting synaptic current can be described as (Destexhe et al 1998):

$$I_{syn} = \bar{g}_{syn} r(t)(V_{syn}(t) - E_{syn}) \tag{3}$$

where $\bar{g}_{syn}$ is the maximal synaptic conductance, $V_{syn}(t)$ is the postsynaptic potential and E_{syn} is the synaptic reversal potential. For the simulations, all synaptic parameters were set to values that gave the best fit to whole-cell recorded synaptic currents (see Destexhe et al 1998). Parameters for AMPA synapses: $\alpha = 1.1 \times 10^{-6} \, M^{-1} s^{-1}$, $\beta = 190 \, s^{-1}$, and $E_{AMPA} = 0 \, mV$. Parameters for GABA$_A$ receptors: $\alpha = 5 \times 10^{-6} \, M^{-1} s^{-1}$, $\beta = 180 \, s^{-1}$, and $E_{GABAA} = -80 \, mV$. Synaptic plasticity was simulated by adapting the maximal synaptic conductance $\bar{g}_{AMPA}$ for recurrent excitatory synapses onto excitatory neurons and GABAergic interneurons according to the learning mechanism described in the text. Inhibitory synapses were not adapted since evidence is currently lacking for their plasticity. We therefore used the following fixed values for $\bar{g}_{GABAA}$ (in μS): 0.04 for Fig. 4, 0.05 for mutual inhibition between the two chains and 0.016 for recurrent inhibitory connections within a chain for the simulations in Fig. 5.

Synaptic plasticity was simulated by changing maximal synaptic conductance $\bar{g}_{AMPA}$ by an amount equal to $\Delta\bar{g}_{AMPA} = \alpha(P_{t+\Delta t} - P_{t-\Delta t})$ for each presynaptic spike at time t, where P_t denotes the postsynaptic membrane potential at time t. The conductance was adapted whenever the absolute value of $\bar{g}_{AMPA}$ exceeded $10 \, mV$ with a gain α in the range 0.02–0.03 $\mu S/V$. The maximum value attainable by a synaptic conductance was set equal to $0.03 \, \mu S$. Note that the learning rule above differs from the pure TD(0) learning rule in that it depends on postsynaptic activity Δt ms in the future as well as Δt ms in the past whereas the TD(0) rule depends on future and current postsynaptic activity (see Equation 1). This phenomenological model of synaptic plasticity is consistent with known biophysical mechanisms such as calcium-dependent and NMDA receptor-dependent induction of long-term potentiation (LTP) and depression (LTD) (see Senn 1997, Egelman & Montague 1998, for possible biophysical implementations).

Acknowledgments

This work was supported by the Alfred P. Sloan Foundation and Howard Hughes Medical Institute. We thank Margaret Livingstone, Dmitri Chklovskii, David Eagleman, and Christian Wehrhahn for comments and suggestions.

References

Abbott LF, Blum KI 1996 Functional significance of long-term potentiation for sequence learning and prediction. Cereb Cortex 6:406–416

Abbott LF, Varela JA, Sen K, Nelson SB 1997 Synaptic depression and cortical gain control. Science 275:220–224

Barlow H 1998 Cerebral predictions. Perception 27:885–888

Bell CC, Han VZ, Sugawara Y, Grant K 1997 Synaptic plasticity in a cerebellum-like structure depends on temporal order. Nature 387:278–281

Ben-Yishai R, Bar-Or RL, Sompolinsky H 1995 Theory of orientation tuning in visual cortex. Proc Natl Acad Sci USA 92:3844–3848

Berry MJ 2nd, Brivanlou IH, Jordan TA, Meister M 1999 Anticipation of moving stimuli by the retina. Nature 398:334–338

Bi GQ, Poo MM 1998 Synaptic modifications in cultured hippocampal neurons: dependence on spike timing, synaptic strength, and postsynaptic cell type. J Neurosci 18:10464–10472

Braitenberg V, Schüz AA 1991 Anatomy of the cortex. Springer-Verlag, Berlin

Chance FS, Nelson SB, Abbott LF 1999 Complex cells as cortically amplified simple cells. Nat Neurosci 2:277–282

Creutzfeldt OD 1977 Generality of the functional structure of the neocortex. Naturwissenschaften 64:507–517

Daugman JG, Downing CJ 1995 Demodulation, predictive coding, and spatial vision. J Opt Soc Am A 12:641–660 (erratum: 1995 J Opt Soc Am A 12:2077)

Destexhe A, Mainen ZF, Sejnowski TJ 1998 Kinetic models of synaptic transmission. In: Koch C, Segev I (eds) Methods in neuronal modeling, 2nd edn: From ions to networks. MIT Press, Cambridge, MA, p 1–25

Douglas RJ, Martin KAC, Whitteridge D 1989 A canonical microcircuit for neocortex. Neural Computation 1:480–488

Douglas RJ, Koch C, Mahowald M, Martin KA, Suarez HH 1995 Recurrent excitation in neocortical circuits. Science 269:981–985

Egelman DM, Montague PR 1998 Computational properties of peri-dendritic calcium fluctuations. J Neurosci 18:8580–8589

Elman JL 1990 Finding structure in time. Cognit Sci 14:179–211

Feidler JC, Saul AB, Murthy A, Humphrey AL 1997 Hebbian learning and the development of direction selectivity: the role of geniculate response timings. Network: Comput Neural Syst 8:195–214

Gerstner W, Kempter R, van Hemmen JL, Wagner H 1996 A neuronal learning rule for sub-millisecond temporal coding. Nature 383:76–81

Hines M 1993 In: Eeckman FH, (ed) Neural systems: analysis and modeling. Kluwer, Boston, MA, p 127–136

Hubel DH, Wiesel TN 1974 Uniformity of monkey striate cortex: a parallel relationship between field size, scatter, and magnification factor. J Comp Neurol 158:295–305

Humphrey AL, Saul AB 1998 Strobe rearing reduces direction selectivity in area 17 by altering spatiotemporal receptive-field structure. J Neurophysiol 80:2991–3004

Jordan MI 1986 Attractor dynamics and parallelism in a connectionist sequential machine. In: Proceedings of the Eighth Annual Conference of the Cognitive Science Society, Amherst, MA, August 1986. Lawrence Erlbaum, Hillsdale, NJ, p 531–546

Kalman RE 1960 A new approach to linear filtering and prediction theory. Trans ASME J Basic Eng 82:35–45

Koch C 1999 Biophysics of computation: information processing in single neurons. Oxford University Press, New York

Levy WB, Steward O 1983 Temporal contiguity requirements for long-term associative potentiation/depression in the hippocampus. Neuroscience 8:791–797

Livingstone MS 1998 Mechanisms of direction selectivity in macaque V1. Neuron 20:509–526

Mainen ZF, Sejnowski TJ 1996 Influence of dendritic structure on firing pattern in model neocortical neurons. Nature 382:363–386

Markram H, Lübke J, Frotscher M, Sakmann B 1997 Regulation of synaptic efficacy by coincidence of postsynaptic APs and EPSPs. Science 275:213–215

Mehta MR, Barnes CA, McNaughton BL 1997 Experience-dependent, asymmetric expansion of hippocampal place fields. Proc Natl Acad Sci USA 94:8918–8921

Minai AA, Levy WB 1993 Sequence learning in a single trial. In: Proceedings of the 1993 world congress on neural networks, July 1993, Portland OR, II, Lawrence Erlbaum, Hillsdale, NJ, p 505–508

Mineiro P, Zipser D 1998 Analysis of direction selectivity arising from recurrent cortical interactions. Neural Comput 10:353–371

Montague PR, Sejnowski TJ 1994 The predictive brain: temporal coincidence and temporal order in synaptic learning mechanisms. Learn Mem (Cold Spring Harb)1:1

Movshon JA, Van Sluyters RC 1981 Visual neural development. Annu Rev Psychol 32:477–522

Mumford D 1994 Neuronal architectures for pattern-theoretic problems. In: Koch C, Davis JL (eds) Large-scale neuronal theories of the brain. MIT Press, Cambridge, MA p 125–152

Nelson S, Toth L, Sheth B, Sur M 1994 Orientation selectivity of cortical neurons during intracellular blockade of inhibition. Science 265:774–777

Olshausen BA, Field DJ 1997 Sparse coding with an overcomplete basis set: a strategy employed by V1? Vision Res 37:3311–3325

Rao RPN 1999 An optimal estimation approach to visual perception and learning. Vision Res 39:1963–1989

Rao RPN, Ballard DH 1997 Dynamic model of visual recognition predicts neural response properties in the visual cortex. Neural Comp 9:721–763

Rao RPN, Ballard DH 1999 Predictive coding in the visual cortex: a functional interpretation of some extra-classical receptive-field effects. Nat Neurosci 2:79–87

Scheinerman ER 1995 Invitation to dynamical systems. Prentice Hall, New Jersey

Schultz W, Dayan P, Montague PR 1997 A neural substrate of prediction and reward. Science 275:1593–1599

Sejnowski TJ 1986 Open questions about computation in cerebral cortex. In: McClelland JL, Rumelhart DE (eds) Parallel distributed processing, vol 2: Explorations in the microstructure of cognition. MIT Press, Cambridge, MA, p 372–389

Senn W, Tsodyks M, Markram, H 1997 An algorithm for synaptic modification based on exact timing of pre- and post-synaptic action potentials. In: Gerstner W, Germond A, Hasler M, Nicoud JD (eds) Artificial neural networks—ICANN'97, proceedings of the 7th international conference, Lausanne, Switzerland. Lect Notes Comput Sci 1327:121–126

Sillito AM 1975 The contribution of inhibitory mechanisms to the receptive field properties of neurones in the striate cortex of the cat. J Physiol (Lond) 250:305–329

Softky WR 1996 Unsupervised pixel-prediction. In: Touretzky DS, Mozer MC, Hasselmo ME (eds) Advances in neural information processing systems 8: proceedings of the 1995 Conference, MIT Press, Cambridge, MA p 809–815

Somers DC, Nelson SB, Sur M 1995 An emergent model of orientation selectivity in cat visual cortical simple cells. J Neurosci 15:5448–5465

Stuart GJ, Sakmann B 1994 Active propagation of somatic action potentials into neocortical pyramidal cell dendrites. Nature 367:69–72

Suarez H, Koch C, Douglas R 1995 Modeling direction selectivity of simple cells in striate visual cortex within the framework of the canonical microcircuit. J Neurosci 15:6700–6719

Sutton RS 1988 Learning to predict by the method of temporal differences. Machine Learning 3:9–44

Thomson AM, Deuchars J 1994 Temporal and spatial properties of local circuits in neocortex. Trends Neurosci 17:119–126

Tootell RB, Switkes E, Silverman MS, Hamilton SL 1988 Functional anatomy of macaque striate cortex. II. Retinotopic organization. J Neurosci 8:1531–1568

Tsodyks MV, Markram H 1997 The neural code between neocortical pyramidal neurons depends on neurotransmitter release probability. Proc Natl Acad Sci USA 94:719–723 (erratum: 1997 Proc Natl Acad Sci USA 94:5495)

van Essen DC 1985 Functional organization of primate visual cortex. In: Peters A, Jones EG, (eds) Cerebral cortex, vol 3: Visual cortex. Plenum Press, New York p 259–329

Wimbauer S, Wenisch OG, Miller KD, van Hemmen JL 1997 Development of spatiotemporal receptive fields of simple cells: I. Model formulation. Biol Cybern 77:453–461

Zhang LI, Tao HW, Holt CE, Harris WA, Poo M 1998 A critical window for cooperation and competition among developing retinotectal synapses. Nature 395:37–44

DISCUSSION

Laughlin: Your model is very elegant, but it is what Mittlestaedt, the famous German physiologist, described as an 'implicator' model. It is a hypothesis that will work. The difficulty in modelling and understanding the cortex is turning an 'implicator' model into what he called an 'explicator' model, where you actually show that the thing that you imply is actually implemented in the system and works. How would you go from testing your model within the cortex to seeing whether it really is operating? We can see that in principle it could operate, but there may be many other models that would produce the same end result.

Sejnowski: The model I presented should be compared with other possible models, which can be experimentally distinguished. An alternative model for direction selectivity is a feed-forward model in which one of two nearby stimuli is delayed or low-pass filtered. The two signals can then be compared at two different times. This is a completely different mechanism for detecting motion from the model I presented, which depends on time delays within the cortex. I think it is valuable as an alternative. It is useful to know what the possibilities are before you begin, so when you do an experiment you can sort out which is more likely to be the case. How would you actually put this to a test? We did, in the sense that when we worked on the model we didn't know what the actual fields looked like in the monkey cortex. The model had properties that are the signature so-to-speak of this type of network, and these turned out to be exactly the properties of neurons recorded by Margaret Livingstone in monkey visual cortex. She went on to do some other experiments in which, instead of flashing just one bar, she flashed a bar at one location, and then after a short delay another bar at another location. She varied the relative times and the relative positions. This produced a much more complex pattern of response. When we did that same experiment in our model, it also looked very similar in terms of the patterns of suppression and activation. As

Klaus Prank suggested, one way to test the model is to vary the stimulus, making it more complex, looking at it under more widely different conditions to see whether or not you can break the model. So far, all the tests are working out qualitatively in terms of the actual patterns of activity that are seen *in vivo*. You would like to go further, I know. You would like to ask whether there is a way to go in and look at the synapses. We are suggesting something radical, that there is a temporal asymmetry: there should be groups of neurons connected together in a chain that have very asymmetric connectivity. There will be anatomical techniques some day to test this. For example, you can inject one cell with a virus that jumps across synapses. It would be great to inject virus into a direction-selective cell to see whether it jumps into the neurons in exactly the right direction. I think it will be possible to test the model some day in a way that will satisfy Mittlestaedt.

Laughlin: You raised the Reichardt model. The big distinction between your model and the Reichardt model is that the latter is hard-wired. Your model is plastic and is based upon coincidences. If you fed the cortex some rather bizarre coincidences, it might learn to anticipate those rather than the simpler pattern of motion.

Sejnowski: There is an interesting story here, which again goes back to Hubel and Wiesl. If you look at the properties of neurons in monkeys, you discover that there are already cells that are selective for orientation, and some that are selective for direction of motion. Initially, the tuning is very weak and the number of cells that are selective is relatively small. Over a critical period during the first few months of life, interaction with the world will sharpen up the tuning of these cells and more will become tuned. In the case of binocular vision, cells that are selective for disparity between the two eyes will develop. If for some reason one eye is strabismic, the two eye images cannot be fused in the brain and the cortical cells never develop binocular input and there will be no stereo vision. We therefore know that there is this period of plasticity in the cortex where correlated inputs are needed. When kittens are brought up in a stroboscopic environment, where there is no continuous motion, the number of cells which respond to direction of motion goes down.

Aertsen: The way I understand it, the rule that you are applying in your model network should, in addition to selectivity for movement direction, also give rise to selectivity for magnitude of stimulus velocity. Do you indeed observe such selectivity in your model and, if so, does it match the properties of the biological cortex?

Sejnowski: There is a preferred velocity, which has to do with the distance and the time delays within the cortex itself. We compared our model with the cortex in terms of the distances between neurons and the actual delays that are observed in the cortex. Because the cortex has a magnification factor that greatly expands the representation of the foveal region, the sensitivity to speed tuning in the fovea has a peak that is much lower in velocity than in the periphery. We were able to match those numbers very well.

Noble: Does this lead you to make some predictions about the tiniest interval that the cortex might be able to distinguish in certain circumstances, and how that would depend on the total number of neurons in a net? My intuition is that the way in which we may come to detect extremely tiny intervals may have to do with the total number of neurons, and this is an explanation then for why you need so many. As an example of this, can we do something similar in relation to auditory signals? The suggestion I have comes from an experiment I did on myself at the age of 40, when I started to learn to play guitar. The first thing my tutor did was to ask me to tune the instrument. The first way I did it, having never tuned a musical instrument before, was to listen for beats. He said this was not the way to do it, and that it was possible to tell immediately. I told him that this was impossible, at least for me at that time, but after about five months I found I no longer had to listen for beats. I have been trying to do the sums on my sheet of paper. If you are telling the difference between 500 cycles per second and 505, waiting for a beat takes quite a long time (at least 200 ms). If you know virtually immediately the difference in timing this has to be detected within a millisecond or so. It is certainly down to a tiny interval. It seems to be like immediately identifying a colour. What I am pressing you for is, could your modelling answer the question of how big a net is needed to detect such very small time differences?

Sejnowski: There are two ways of answering that. One of them would be to look at the smallest displacement that is needed in order to detect motion. This depends on the distance and time. There is a phenomenon known as 'apparent motion' that occurs when one stimulus goes off and a nearby one turns on. We are capable of detecting whether two visual stimuli are simultaneous to within a few milliseconds.

Laughlin: I would say that auditory discrimination is about 50 μs and visual is 300 μs. Just to sharpen up people's appreciation of the nervous system, in owls that need to establish the direction of sound very well, it is below 10 μs, and in electric fish which can establish phase differences extremely well it is below 1 μs. The consensus is that this resolution is achieved by some sort of population code, where you have an array of neurons, each of which is relatively broadly tuned to a unique delay, and then you look for the peak of activity in this array.

Sejnowski: There is another level of precision hyperacuity. If you have two lines that are displaced by a small amount, we can detect this down into the arc second range, despite the fact that rods are about 30 arcseconds across. Again, it is thought that this is due to some population code, although no one understands exactly how it is done.

Fields: Have you preformed any experiments or simulations for the cellular mechanism for the back-propagation effect on plasticity?

Sejnowski: Yes. In the model I showed you this is hardwired in the sense that we took what was there and automatically changed the synapse when it was within the right direction. This learning algorithm can be summarized as follows:

$$\Delta W \alpha (V_t - V_{t-1}) P_t$$

where W is the synaptic strength, V is voltage in the postsynaptic cell and P is the presynaptic signal. This is Hebbian in the sense that in order for the plasticity to take place, you have to have a presynaptic signal present at the same time as a postsynaptic signal. This temporal difference equation yields a temporally asymmetric synaptic learning rule. The insight is that during the back-propagating action potential there is a period during which the voltage is increasing and there is a period in which it is decreasing. If you look at this temporal difference, it is basically a true discrete form of a time derivative. This means that during the rising period of the back-propagating action potential, this difference is positive, which means that you increase the strength of that synapse, and during the falling period, when the spike occurs before the EPSP, the slope of the back-propagating potential is negative and therefore you decrease the strength of the synapse. This is the origin of the concept of temporal difference learning rule in neural networks. You were asking a question about the biophysical mechanisms. During the back-propagating action potential, Ca^{2+} channels open in the spine and Ca^{2+} will enter the spine. If glutamate is already bound to the NMDA receptor, even more Ca^{2+} enters the depolarized spine, binding to Ca^{2+}-binding proteins with a range of off-time constants. Putting all the kinetics and all of the numbers of channels and time courses into the model, we have been able to reproduce this temporal difference learning algorithm. The key to understanding why it works was given by Michael Berridge earlier in the meeting: the Ca^{2+} buffer has to become saturated before free Ca^{2+} can activate the LTP machinery. If you just have the EPSP by itself, it never activates and you'll never get enough free Ca^{2+}. But if, just after the NMDA channel opens, you have the back-propagating action potential coming in, this will allow enough Ca^{2+} to come in for a long enough time to be able to bind to the buffers and therefore allow saturation to occur. Activation of the calmodulin and Ca^{2+}–calmodulin-activated protein kinase type II (CaMKII) then leads to a biochemical cascade leading to LTP. We are beginning to see how to implement, with molecular mechanisms, an abstract learning rule, and then on the basis of this learning rule fed into a network we can see how populations of neurons can produce interesting computation properties in the cortex. There are three levels of modelling here: biophysical, modelling single neurons using a temporal difference rule, and the level of the whole network.

Iyengar: I have been wondering about how the relatively slow reactions that we study fit into what you are doing. Given the rate at which a back-propagation has to work, it has to come fast enough so that CaMKII is not completely dephosphorylated.

Sejnowski: This is another key insight: the duration as well as the magnitude of the Ca^{2+} entry must match the kinetics of CaMK phosphorylation.

Iyengar: You need to go back and activate the phosphatases in such a way that you don't drive the system completely down.

Sejnowski: Activation of the calmodulin-dependent phosphatase, calcineurin, may indeed dephosphorylate CaMKII for prolonged, intermediate levels of Ca^{2+} entry. The balance between LTD and LTP depends on a kinetic race. However, these details don't reveal what computation is going on in the dendrite. The biochemistry is embedded in a much larger network of reactions, which may include gene regulation: from neural networks to gene networks. Someone earlier suggested that different models or theories were needed at each level, and that they shouldn't talk to each other. These links between levels are needed.

Final discussion

Noble: I want us to return to a question that has been haunting this meeting for some time. Can we make the statement in relation to levels that the best models are those that span at least three levels? One can see a reason for this. There will be a level at which you are purely descriptive. (It is very interesting that Hodgkin and Huxley's 1952 paper was entitled a 'description' of membrane current, not a 'model' of membrane current: they never said that they were building a theory.) There is a second level at which a model tries to integrate, and a third level at which it tries to predict. As a general rule, one could say that good models must span three levels.

Sejnowski: The same mathematical model can serve all those different functions. It can first describe the data. Then it suggests a theory that will go beyond the actual data, which has to do with mechanisms, which can be tested.

Berridge: I have a philosophical comment. When do people accept that a project, especially one dealing with modelling, is complete? For example, we talk about the genome project being completed. In the case of modelling the heart, is there any way that we can conceive that there will be a clearly defined endpoint, or is this going to go on for ever?

Brenner: It is asymptotic. Even the genome project is. W. C. Fields once said that for a man who falls off a 300 ft cliff, it is only the last inch that really hurts. I think science is asymptotic, and I also think we change the conditions that we accept for understanding. That is why so many problems just disappear, because in one period we formulate the problems in terms that we think will give us understanding, and then the whole basis of understanding changes. People look back and are puzzled by some of the questions posed in the past. I can remember a period in developmental biology when a crucial question was whether differentiation was a process or a state. This was discussed endlessly but, in hindsight, it seems a crazy question and that problem has certainly vanished. It is a good thing for science, because we develop a new picture with every step of advance. I am reminded of the story about the man who falls off a tall building. On the way down someone at the window sees him go past and asks him how he is. He says, 'So far it's OK'.

Dolmetsch: I would like to ask the opposite question, which is when is it appropriate to start modelling? Many experimental biologists feel that models are often made prematurely, and that they really don't contribute anything until most of the components in a particular threshold are known. Is there some way of determining when that threshold is reached, or should we always make a model because it is a good exercise?

Brenner: If you can make a model, then it is the right time.

Noble: I have a slightly different version of the answer to that, which is that it has to be when you think there is the possibility of answering a question. I was recently asked by someone writing a review article, what was the purpose of the model I developed with Don Hilgemann about 15 years ago, when we first looked at trying to model Ca^{2+} dynamics inside the cell, together with Ca^{2+} buffers represented (Hilgemann & Noble 1987). I compiled a list of six questions that we were asking at the time which we thought needed to be answered. Interestingly, no one asks those questions any longer. One of them, for example, was 'is the Na^+/Ca^{2+} exchanger electrogenic or not?' In its original description this exchanger was neutral. When we were first modelling we had to determine whether it mattered to the Ca^{2+} dynamic system that we were building. We found that it did matter. My answer to your question would therefore be that the right time to model is when you think you can answer a question.

Sejnowski: There is another way to answer the question, which has to do with the need to think clearly about a problem, which raises the issue of complexity. It may be that there are not enough data yet to answer the question, but you don't even know what the question is. In other words, you are dealing with things that are too poorly understood. This is what often happens in the brain. Having explored a model, even in the absence of all the details, we can begin to formulate the question in a way that begins to make sense.

Iyengar: There are some things that we have found in biochemical modelling that help us understand connections mechanistically. We always know of lots of pairwise interactions that result in large models. Some of these models are really helping us tease out which interactions matter and which don't within the time frames of the processes we study. In the model that I published, I coupled MAP kinase activation to cAMP through PKC regulation of adenylate cyclase 2 for regulating the CaMK activity. We have done experiments now where we have shown nicely that MAP kinase activation can regulate CaMK phosphorylation, but there seems to be no role of cAMP in that. This was a connection we made that we thought was the most reasonable on the basis of all the pairwise interaction data. Now we can go back and say we have these two endpoints, so how are the interconnecting mechanisms set up? This would not have come out of thinking about the individual molecules.

Noble: I'd like to react once again to Ricardo Dolmetsch's point, because I think this also raises a general issue that we should tackle: is there in this respect a serious difference between the engineering and physical sciences and the biological sciences? What I am also detecting in Ricardo's question is that there have of course been many early attempts at biological modelling that have been spectacular failures. My reading of the situation is that in the physical sciences this is OK. It has been accepted that there is a theoretical branch to the subject and an experimental part. It is the function of most theories ultimately to be wrong. In the biological sciences we don't seem to have had this tradition, although it may be emerging. Faced with the extreme complexity of biological systems, perhaps we have been super-sceptical about the use of theory.

Iyengar: I would put many scientists who experimentally study the cell cycle in that category. They do not favour modelling because they think it produces no new insight. Many of these models have been somewhat accounting sort of models, where they just put into perspective what is already known rather than push the envelope to try to predict something new.

Sejnowski: The reason why models are effective in physics, in part is that the modellers collaborate with people doing the experiments, and they discover that it gives them an edge on their competition. In biology, this is just beginning to become part of the culture.

Aertsen: One of the critical questions about whether a model will be helpful or not has to do with the number of free parameters in it. As you all know, five parameters make an elephant, so we should try to make it fewer than that.

Noble: I like the point you are making, but I'd like to add a caveat. When we are dealing with systems that have so much detail in them, you are not going to count the fact that you rely on vast databases as having free parameters. Given the complexity of what we are trying to deal with and the huge databases on which we increasingly rely in biological modelling, we are going to have to accept that modelling will be done with huge numbers of parameters. Which of these are free and which are fixed is going to be a matter of debate, to some extent.

Segel: You may be able to make an elephant with five parameters, but you can't make several different animals. In my experience, if you try to get semi-quantitative agreement with several different experiments it is very hard to do. Secondly, concerning the cell cycle, although the models may not interest the experimenters who are trying to get more detail, they definitely have a value. For example, they can explain the systemic nature of how a checkpoint works. There is a receptor which monitors a system property. But this receptor has to lead to the stopping of the cell cycle if something doesn't work. Then, if the blockage is relieved, the cell cycle has to go on as it did before. The basic underlying idea of how this works is described in some of the nice cell cycle models. Even though it may not influence what the experimenters are doing at the moment, such conceptual clarifications are undoubtedly of value (see for example Tyson et al 1996).

Brenner: In the early days of the cell cycle research, before we knew so much about the machinery, there were two models. There was the clock model, which said that the cycle starts, you go to the next position and something happens and so on, a bit like a set of dominos. And then there were people in dynamical system theory who said it was a limit cycle that can be described in terms of chemical reactions. One knew that if it was a domino model, there would be nothing for it but to go and find all the components, which is what they are busy doing in cell cycle research. It isn't that we haven't had models. One just has to think back to Mitchell's theory about energy coupling in oxidative phosphorylation. Biochemists were fixed on the thought that there would be chemical intermediates, and when Mitchell proposed his theory of concentration gradients it was considered to be nonsense. But it is the one that actually is true, and biochemists had to get used to thinking in different terms.

Sejnowski: I gave another example earlier, of Hebb. He guided a lot of early physiology. It never occurred to the physiologists when they first discovered LTP that one should look to see whether it was necessary for the postsynaptic cell to be depolarized during the stimulus. It wasn't at the top of the agenda. But one of the physiologists, Tom Brown, who had read Hebb, was inspired to hyperpolarize the cells and show that this could block LTP. There are now temporally asymmetric versions of Hebbian synapses. This is a case where progress has been made with a theory that involved conceptual frameworks and not necessarily equations with many parameters.

Brenner: I was going to take up the question that Denis Noble raised, that we have to get equations so we can understand each other. Well, you'll have to count me out, because I don't speak that language! What is happening in biology is a change of the conceptual framework in which we are operating. I wrote some time ago that I thought we were generating new strange names, such as computational biology, to conceal what we really want to call it, and that is 'theoretical biology'. This has such a bad name that people are reluctant to use it. I noticed at our meeting that when people have been asked a question about theory, they have prefaced it with the remark, 'I am going to ask a philosophical question'. But these are really theoretical or conceptual questions.

Pozzan: I have a philosophical question. What is the difference between a theory and a model?

Brenner: The term 'model' used to refer to a provisional theory, just as a 'paradigm' used to mean an example. Now, of course, 'paradigm' has come to mean something different, with excellence and uniqueness attached to it. The word model came in with molecular biology. They tried to provide a theory but called it a model because it was based on componentry.

Sejnowski: The structure of DNA was a literal model, made of bits of wood and metal, but inherent in the model is a theory. It was both.

Brenner: People also proposed models of DNA replication, but I think they meant they had a biochemical theory of how it might work in the real world.

Berridge: Denis Noble, what do you call your work on the heart? Is it a model with some theoretical aspects to it?

Noble: I am trying to think carefully in the light of Sydney Brenner's remarks, and that question is precisely what I am asking myself. It seems to me that some aspects of modelling do not involve new theory. You are trying to build a jigsaw, which you want to get as accurate as you can. At this stage in your modelling work, you are not operating with any new theory, you are just filling out the detail. I identified in the discussion earlier on about the question about whether the Na^+/Ca^{2+} model was or was not electrogenic a point at which a model in that stage of development of our ideas about Ca^{2+} handling had to make a hypothesis. It had to say, 'I think this has to be electrogenic and I am now going to build a model to show why this may be so'. By any criteria, you have to say that this is a theory. There is an iterative process here (Noble & Rudy 2001). There are stages of modelling that require that you put forward new hypotheses, but as with experimental work there are also great long tracts of time when what we are doing is largely filling in.

Brenner: People began to model photosynthesis and some thought that electron tunnelling would be required to explain certain phenomena. Electron tunnelling made the prediction that certain processes would go on in the thylakoid membranes at the temperature of liquid nitrogen, which of course is not very biological. Then an experiment was done, that showed that electron tunnelling was found in the membranes and that it could take place at normal temperatures. Once electron tunnelling was found to apply to photosynthesis, where we have unique structures, it began to be applied to everything. There are always people who will take physical theories and try to embed them in models in other fields: some are right and some are wrong. But if you are asking about the enveloping theory, I think it is going to have the flavour of a computational theory. We have representations of things, they interact and they produce something else. This is a sort of guiding scheme in which I can embed my thinking: in crude terms, I start with the data bank the genome provides, from which I have to proceed.

Sejnowski: 'Evolvability' will be an important constraint.

Iyengar: There are also context-dependent functions: molecules function in different ways in different contexts.

Laughlin: Talking about all these theories and saying that we need a theoretical biology rather makes us lose sight of the problem. Going back to what Denis Noble said about why people in physics are allowed to make mistakes with their theories, I think this is because the physical systems are so constrained that the number of possible theories is rather small. The likelihood of getting the right one is, therefore, reasonable. We are working with complex systems, and this means that by definition there are many alternative models which could explain the phenomena. In the past, biologists have been sceptical of theorists because they have simply wandered off into this space of all possible models and have got totally lost.

Sejnowksi: I can assure you that there were a lot of theories in physics that were wrong, too. Models can help you sort out the right ones because they allow distinct comparison with experiment.

Laughlin: The scale on which you can be wrong in biology is much greater. The two things we have to guide us and keep us from getting lost are the data and concepts.

Brenner: I would like to provide two examples in which we can see how a general theoretical framework involving the symmetry of protein–protein interactions can help us in thinking about the evolution of regulation. When the model of allosteric inhibition was introduced by Monod, the idea was that the allosteric sites were not the same as the active site, and that in proteins with multiple subunits, there were inhibitory subunits for the allosteric sites and regulation took place by interaction through the proteins. The actual example was haemoglobin, a tetramer. This led to the idea that if a protein was not a monomer, the oligomeric state must exist for purposes of regulation. We know dozens of enzymes that are oligomers with no evidence at all that regulation is involved. Thus we conclude that making dimers must be quite common; sometimes it is used in regulation and sometimes not. We need a theory for why proteins exist in this form. I want to give you one that was formulated by Francis Crick and myself when we were interested in this problem. Let us assume that we have a set of proteins in a cell subjected to random mutation, and that some mutations change the charge distribution on the surface of the protein so that it can interact with itself. It is easy to see that the most probable structure generated is an infinite helix, which may precipitate and be disadvantageous to the cell. There would be selection to remove this state, and one way would be to reverse the original mutation, a relatively rare event. Another would be a compensatory mutation, such as additional charges which occur around a dyad axis, because then the interaction becomes closed and does not propagate.

Sejnowski: So a dimer is a cell's best friend!

Brenner: Yes, and the argument continues that the same might have happened to dimers, which will create another asymmetry axis and make a tetramer. Finally, we can consider the case of aspartate transcarbamylase. It has a regulatory subunit that binds cytosine triphosphate (CTP), the end product of the pathway. How can we explain how this recognition site arose? It is unlikely that we evolved a site for this on the surface of a protein. Perhaps there existed somewhere else in the cell another enzyme with a recognition site for CTP. Random mutations might have brought these two proteins together, and if their interaction was of advantage to the cell, then there would have been selective pressure to improve it. The original function could be retained eventually by a gene duplication, two versions could be generated, one of which was the enzyme and the other the allosteric subunit. This is a theory which gives a plausible account of how a certain state may have arisen by a series of small steps and does not require one complicated jump.

Sejnowski: I'd like to draw this meeting to a close. We have just begun to get a sense for the issues here: I see this meeting as a positive beginning. We are reaching a stage with the science of complexity where theoretical thinking is beginning to be helpful. There isn't going to be any single paradigm or prescription. We are taking our constraints from wherever they come and getting our insights from many different places. This is exactly the way it should be. I think the way we will converge on a better understanding is by intersecting constraints from many different levels. It is not a question of which is the right model, but whether or not a model is useful. Models are tools to facilitate progress, and like any experimental tool they have flaws. We are seeing this convergence of constraints happening from the genetic direction with much better cataloguing of what is there. We are also seeing this with physiological techniques such as Ca^{2+} and cAMP imaging that make explicit the spatial heterogeneity in the cell. We are dealing here not with a soup, but with a highly structured cellular milieu in which there are ordered organelles and proteins that are interacting in a way that produces a useful function. There are a finite number of molecular machines that the cell uses in order to produce all the functions we see. I am also very impressed with the convergence going on between the biochemists and neurobiologists. I began this meeting by quoting Dobzhansky, who said that nothing in biology makes sense except in the light of evolution. This brings us full circle to the close of this meeting. I have had a wonderful time and would like to thank you all for your contributions.

References

Hilgemann DW, Noble D 1987 Excitation–contraction coupling and extracellular calcium transients in rabbit atrium: reconstruction of basic cellular mechanisms. Phil Trans Proc Roy Soc B Biol Sci 230:163–205

Noble D, Rudy Y 2001 Models of cardiac ventricular action potentials: iterative interaction between experiment and simulation. Phil Trans Roy Soc B Biol Sci, in press

Tyson JJ, Novak B, Odell GM, Chen K, Thron CD 1996 Chemical kinetic theory: understanding cell cycle regulation. Trends Biochem Sci 21:89–96

Index of contributors

*Non-participating co-authors are indicated by asterisks. Entries in **bold** indicate papers; other entries refer to discussion contributions.*

241

Subject index